Springer Proceedings in Energy

The series Springer Proceedings in Energy covers a broad range of multidisciplinary subjects in those research fields closely related to present and future forms of energy as a resource for human societies. Typically based on material presented at conferences, workshops and similar scientific meetings, volumes published in this series will constitute comprehensive state-of-the-art references on energy-related science and technology studies. The subjects of these conferences will fall typically within these broad categories:

- Energy Efficiency
- Fossil Fuels
- Nuclear Energy
- Policy, Economics, Management & Transport
- Renewable and Green Energy
- Systems, Storage and Harvesting

Materials for Energy eBooks Volumes in the Springer Proceedings in Energy will be available online in the world's most extensive eBook collection, as part of the Springer Energy eBook Collection. Please send your proposals/inquiry to Dr. Loyola DSilva, Senior Publishing Editor, Springer (loyola.dsilva@springer.com)

More information about this series at http://www.springer.com/series/13370

Paolo Bertoldi
Editor

Improving Energy Efficiency in Commercial Buildings and Smart Communities

Proceedings of the 10th International Conference IEECB&SC'18

Springer

Editor
Paolo Bertoldi
European Commission Joint Research Centre
Institute for Energy, Transport and Climate
Ispra, Varese, Italy

ISSN 2352-2534 ISSN 2352-2542 (electronic)
Springer Proceedings in Energy
ISBN 978-3-030-31458-3 ISBN 978-3-030-31459-0 (eBook)
https://doi.org/10.1007/978-3-030-31459-0

This Springer imprint is published by the registered company Springer Nature Switzerland AG
The registered company address is: Gewerbestrasse 11, 6330 Cham, Switzerland

Introduction

Commercial (i.e., non-residential) buildings, public buildings and urban areas are the fastest growing energy consuming sectors, and keys to where CO2 reduction could be achieved in a cost effective manner. This is mainly due to the growth of commercial and public activities and their associated demand for heating, cooling, ventilation (HVAC), ICT services and lighting.

As a consequence all actors need to take all necessary steps to increase efficiency and implement renewable energies by disseminating good practice, fostering investments and providing technical solutions for the commercial building sector and districts/communities. This includes behavioural changes in the way companies, architects, urban planner and building occupiers invest, design and operate non-residential buildings.

The integration of distributed generation, district heating and cooling and renewable energy sources (RES), demand response and smart grids would enable further CO_2 and energy saving and pave the way for Net Zero Energy Buildings (NZEBs) and Districts (NZEDs). Energy Service Companies (ESCOs), utilities, telecoms and facility management companies offer advanced solutions to monitor, manage and reduce the energy consumption in commercial buildings.

Very often building energy performance can be more cost-effectively optimized at the district or urban level. Therefore the conference has a special track on smart and sustainable districts and communities.

A number of local, regional and national policies and programmes have recently been implemented to achieve a long lasting market transformation, including building codes, energy performance certificates, utilities programmes, energy audits, information and training, emission and energy certificate trading, and financial incentives.

Following the success of the previous IEECB conferences, the Messe Frankfurt and the European Commission DG JRC, organised the 10th *International Conference on Improving Energy Efficiency in Commercial Buildings and Smart Communities (IEECB&SC' 18).*

The IEECB&SC'18 took place on 21–22 March 2018 in Frankfurt, during Light+Building, the world's leading trade fair for lighting and building services technology. The conference brought together all the key players from this sector, including commercial building investors and property managers, academia, building technologies researchers, equipment manufacturers, service providers (ESCOs, utilities, facilities management companies), urban planners and local policy makers, with a view to exchange information, to learn from each other and to network.

In particular the conference attracted property owners, investors, architects, local authorities and urban planners to present and discuss synergies and cooperation in removing existing barriers to energy efficiency, renewable energy and smart buildings and districts.

The IEECB&SC'18 conference attracted high level presentations showing new technologies, techniques, services, policies, programmes and strategies to increase energy efficiency, energy savings and to reduce greenhouse gases emissions in non-residential buildings and district/communities.

Papers by authors covered the following topics:

1. *Lighting, Appliances and Equipment*: technologies (light sources, LEDs, luminaires, control gear, and control systems), day-lighting, Green Lights programmes, lighting quality and energy efficiency, simulation and design tools, commercial refrigeration, cooking and washing, vending machines, lifts, equipment labelling and standards.
2. *Building envelope, passive techniques and HVAC*: low energy cooling techniques, passive cooling and natural ventilation, solar cooling, techniques for low energy fluid movement, heat/cool storage, indoor air quality and energy efficiency, test methods and simulation tools, building and ductwork airtightness, façade technologies (e.g., double skin facades, roofing, etc.), new insulation and phase changing materials.
3. *Examples of advanced/demonstration buildings*: results of new building concepts and smart buildings; successful refurbishment, successful integration renewable energy sources, buildings integrated planning for energy efficiency, Zero-Energy and Positive buildings, office buildings, supermarkets and commercial centres, hospitals and schools, airport & train stations.
4. *Information and communication technology (ICT) equipment and data centres*: data centres design and optimisation, efficient servers, network and storage equipment, the impact of internet on commercial building consumption, data networks, telecom and broadband networks energy efficiency, Energy Star programme for ICT.
5. *Renewable energy sources, distributed electricity and heat generation*: co-generation and poly-generation, micro turbines, heat pumps, fuel cells, biomass boilers and renewable energy sources (solar thermal, PV, etc.), successful PV building integration, building as centre of the smart grid, district heating and cooling, energy communities and co-operatives.
6. *Control Systems, IoT and Building Energy Management Systems (BEMS)*: R&D & technologies, successful implementation, impact on energy consumption and indoor quality.

7. *Energy and facility management, energy services*: continuous commissioning and retro commissioning, energy audits, optimisation of building operation, energy management, role of the energy/facility manager, operation and maintenance, outsourcing of building energy management, education and training of facility managers. Energy service companies (ESCOs), Energy Performance Contracting, financial institutions, public-private partnerships, new financial options, carbon financing, M&V.
8. *Policies and Programmes (local, national or International)*: building codes (new and existing buildings), building certification, code compliance, best practice programmes, energy audits, energy company obligations (EERs, white certificates, etc.), national and local energy efficiency funds, Green Buildings and Energy Star programmes, building rating, building quality labels, voluntary building certification systems, Life Cycle Costing (LCC), programme evaluation, green procurement, building code compliance, national roadmaps for nearly zero-energy buildings, building renovation and cost-optimality, role of buildings and cities in reaching climate targets (e.g. 1.5 °C).
9. *Energy consumption monitoring and benchmarking, Energy Modelling of Building performances*: building/district monitoring campaigns, data analysis and assessment of consumption of specific equipment, assessment of building standby consumption, energy efficiency indicators for buildings, benchmarking, understanding and fixing the disconnect between predicted and measured performance, dynamic simulation methods, software and tools for design of low-energy/zero and positive buildings and building systems, GIS systems.
10. *Demand response*: Demand Response programmes and technologies, dynamic tariffs, results and evaluation, practical implementation in non-residential buildings, impact of real time energy consumption feedback.
11. *Behaviour and barriers to energy efficiency, Investors' motivation and financing*: Marketing and selling energy efficient buildings, costs and benefits analysis including non-energy benefits, market impact of energy performance certificates, corporate social responsibility, value of green buildings, facilitation of the planning process for low energy buildings/districts, interaction between investors, planners, architects, engineers, and users, non-technical barriers efficiency in commercial buildings, analysis of behavioural aspects in the commercial buildings sector and urban areas, and ways to overcome them.
12. *Sustainable and smart communities, districts and cities*: challenges and opportunities with integrating buildings into wider community energy planning; district energy systems; community demand balancing; innovative economic and business models to share risk and benefits across community energy structures; integration of smart building and smart grids; smart cities, integration of community energy planning in urban planning, zero carbon district energy systems, urban strategies for improving energy efficiency in communities, sustainable university campus and labs, impact of urban morphology on energy reduction policies. Urban forms play an important role in achieving energy saving measures, resilience thinking approaches and urban ecology design, cities emissions inventories, sustainable energy and climate action plans.

Contents

Demand Side Management in the Services Sector: Empirical Study on Four European Countries

Ulrich Reiter, Robin Peter, Katharina Wohlfarth, and Martin Jakob

Abstract Demand side management (DSM) is seen as promising, cost-effective measure to cope with high shares of intermittent renewable energy in the electricty grid system. As the regulatory framework in Europe is changing in favour of opening up new market opportunities for such measures, it raises the question which DSM potentials are effectively available. Besides the DSM potential in the industry sector, which is already addressed in many countries, the information on the DSM potentials and market acceptance in the services and residential sector is scarce. In order to properly evaluate such potentials and their impact, quality data is of utmost importance to understand the barriers and drivers for the future market development. Therefore, an empirical study regarding the DSM potential in the services sector is conducted to collect firsthand data from potential DSM users.

In this paper we present the findings of the empirical study, describing the results for the tertiary sector of the following European countries: the UK, Poland, Italy and Switzerland. Our study includes the subsectors retail, wholesale trading, hotels, restaurants, office-type companies (privately held), public administration, public companies and services. The collected data is important and highly necessary as it remains currently unknown which facilities have already been included in DSM-markets and what willingness or readiness is dormant in services companies, to govern over specific facilities. The data-set and the results of the study were collected within the EU REFLEX project [The project REFLEX has received funding from the European Union's Horizon 2020 research and innovation program under grant agreement No 691685 and the Swiss State Secretary for Education, Research and Innovation (SERI)] and will be used as basis for further modelling exercises, to analyse and evaluate the development towards a low-carbon energy system with focus on flexibility options in the EU.

U. Reiter (✉) · R. Peter · M. Jakob
TEP Energy GmbH, Zurich, Switzerland
e-mail: Ulrich.Reiter@tep-energy.ch; Martin.Jakob@tep-energy.ch

K. Wohlfarth
Fraunhofer-Institut für System- und Innovationsforschung ISI, Karlsruhe, Germany
e-mail: Katharina.Wohlfarth@isi.fraunhofer.de

P. Bertoldi (ed.), *Improving Energy Efficiency in Commercial Buildings and Smart Communities*, Springer Proceedings in Energy,
https://doi.org/10.1007/978-3-030-31459-0_1

1 Introduction

With the rise of new, more efficient means to harvest renewable energy and humanity's ever-increasing hunger of electric energy, one of our main concerns has not changed since the upcoming of the use of it; the ability to effectively store electric power.

The scope of the described work was to analyse the current situation of demand-side management (DSM) in major stakeholders in peak energy usage as promising utilizers of such initiation and their potential contribution to implement DSM options in the future.

This paper presents the methods and results of a qualitative stakeholder survey to further determine barriers and thresholds, potentials and drivers, as well as designate a concrete inception of DSM (see also [1]). While the DSM potential from the industry sector is better understood and better technologies have been developed for a facilitated DSM implementation, such as smart grids [2], the expectations and boundary conditions from service sector market participants is widely unknown and therefore untapped. So far, there is an insufficient amount of data available about many areas of the European service sector to estimate the current DSM potential. It remains unknown which facilities have already been included in DSM-markets and what willingness or readiness is dormant in facility operators to govern over specific facilities. Furthermore, it has not been clearly identified which obstacles and restrains are crucial to decide to participate in the DSM-market or not.

In order to empirically address these topics, a survey directly aimed at companies in the service sector was performed, to determine based on the findings whether there is potential to improve taxable loads. The selected stakeholders include subordinates such as, aggregators, network operators, energy providers, technology suppliers and the Federal Network Agency, among others, but also sector agents like trading companies, bureaus and hotels. They were selected across four nations; the UK, Italy, Poland and Switzerland.

Ultimately, this article will describe, firsthand information on the state of the current aptitude of an implementation of DSM into the market and explain the implications. Thus, the focal point is laid on research questions concerning sectors and key aspects of an optimized usage of electric power at peak times. There are several technologies on which DSM would be particularly effective, as they consume, comparatively, large amounts of energy and due to the nature of their functionalities, they are meant to be used for long periods of time without breaks in between. Such technologies comprise air conditioning, cooling and refrigerating and ventilation systems.

DSM can be suitable for two types of clients: on the one hand for bigger companies, which have elevated loads and consumption of electricity. Likewise, larger establishments will have an energy management system (EMS) as well as an energy manager at their disposal. On the other hand, smaller companies with non-process related technologies (heat pumps, etc.) could be aggregated by service companies, thus making available untapped potentials in a cost-effective manner.

However, even companies of large scales may encounter further restrains and obstacles to overcome. Companies generally refrain from change, avoid investments and fear disturbance of work flow and quality.

Furthermore, there will be inhibitory regulatory frameworks such as the German StromNEV (Electricity Network Fee Regulation Ordinance aiming for the avoidance of load fluctuations), incomplete aggregator-models impeding pooling. Also, the market is not accessible without having to undergo complex qualification processes. Overall, additional policy measures are needed to promote DSM in the services sector as described in [3].

To conclude this introduction, currently there are potential cross-sectional technologies at hand, and partial previous experiences as well as facilitating condition are present. Notwithstanding, due to regulatory frames there is barely space to access DSM's profitable potentials. Only time will show the willingness and readiness at an enterprise level, as well as its practical, usable potentials.

2 Methods

In total, 1206 complete data sets were collected from corporations of four service sectors. A set of data counts is complete when a survey participant has answered all posed questions concerning a domicile (location) of a company. At maximum, three domiciles per concern or per international enterprise respectively are permitted to participate (or one domicile of a company respectively).

2.1 Selection of Sample

The focus was set on four specific service sectors to include wholesale/retail, hotels and restaurants, private office-type companies, and public administration. Each contained at least 75 data sets, adding up to 300 data sets at minimum per country (see Table 1 for more details). The following selective criteria have to be paid attention to, when selecting possible corporations from tertiary sector, which are to be surveyed by the contractor:

1. Economic sector/sub-sector.
2. Host country.
3. Size of company (number of employees).

Table 1 Overview of sample size

	Bureau		Public sector	Trade				Hotel, restaurants		Total
Number of enterprises	Private	Public	Health/Education	Retail	Wholesale	Of which Food[a]	Of which Non-food[a]	Hotels	Restaurants	
UK	75	39	36	25	51	(14)	(63)	40	35	301
CH	75	54	22	42	33	(16)	(63)	35	40	301
IT	75	75	1	7	68	(21)	(58)	69	8	303
PL	73	56	21	49	26	(13)	(62)	43	33	301
Total	298	224	80	123	178	(64)	(246)	187	116	1206

[a]Multi-selection possible for specification between food and non-food trade companies
Number of respondents per country and sub-sector

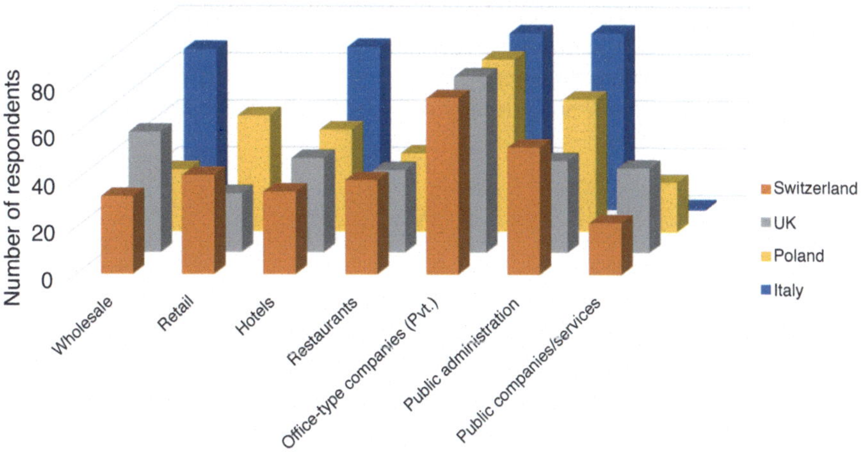

Fig. 1 Number of survey respondents per country and sub-sector

Economic Sector

Based on the country structure of the service companies and the available address data, the distribution of respondents between sub-sectors and countries varies (see Fig. 1). This is of importance to understand and consider differences for calculating country wide DSM potentials and find analogies for estimating DSM potentials of countries not considered in the survey directly.

The selection of sub-sectors is given as follows: Wholesale and retail trade; hotels and restaurants; office type companies with privately owned office-type companies include banking and insurance companies. The public administration encompasses office-type buildings. The public companies and services are composed of healthcare, education and schools, and culture.

Host Country

To access and participate in the DSM market, the two following main conditions need to be fulfilled (among others):

1. The regulator needs to adjust the market framework conditions to allow consumer to participate in demand response (DR) programs.
2. Companies need to fulfil various regulations and technical standards [4] to be eligible to participate in the ancillary services market.

The general set-up for this framework on EU level is defined by the Article 15.8 in the European Energy Efficiency Directive [5]. However, looking at country levels, the definition of such framework conditions varies strongly and often lacks full implementation. To get a grasp on the different market statuses, the empirical study was conducted in countries where the market design is at different development stages. Whereas in Switzerland (CH) and the United Kingdom (UK), market regulations for DR options are already in place [6], other countries such as Poland (PL) are lagging behind or are currently designed as very restrictive such as Italy (IT) (ibid.).

Size of Company

Regarding the size of the company, different approaches are expected to be implemented in terms of energy efficiency and DR. While the specific energy demand per employee and sub-sector varies, and given the limited number of samples possible, only two group-sizes were defined to distinguish between market participants. While service companies with few employees often dominate specific sub-sectors, the overall energy demand and therefore the DSM potential is rather lower (exceptions exist) as compared to large companies with high energy demand.

2.2 *Survey Design*

The questions of the survey can be categorized into four different main energy-related topics:

1. General information on the firm.
2. Energy efficiency.
3. DSM solutions.
4. Decision processes.

First, some general information on the building or the site needed to be collected, in order to be able to put the results into an appropriate context of size and energy demand of the companies. General questions focused on the number of employees, the rented or owned energy reference area, annual electricity consumption and the way the demand is metered and the existing building standard, among others.

Next, the questions focused on the relation of the company towards energy efficiency, such as the companies' commitment to energy audits, planned or implemented energy or environmental management systems or expected investment or refurbishment measures to reduce overall energy demand.

The third topic is focusing on evaluating the technological readiness of companies towards DSM solutions, the acceptance of DSM and the willingness to install the technologies that go alongside with load management (e.g., company allows for DSM, is already participating, what are drivers and hurdles, economic expectance, among others). Depending on the use of DSM options already today or not, participants were asked which technologies they have installed on site, which of them they are already integrating in their DSM contract and which other installations could be used for DSM.

Finally, the questions about the decision processes were raised, to understand the position of the respondent within the firm and which decision levels need to be addressed to realize investments in energy efficiency or DSM.

3 Results

The results of the stakeholders will be explained for all countries in an overview to allow direct comparison between companies' preferences. The most relevant and telling histograms are described and further elaborated upon in the subsequent subchapters. Alongside the descriptive statistics, the technological equipment of each of the four surveyed countries are outlined below and offer a clear insight about how feasible the implementation of DSM is so far and what are potential contributions towards future use of DSM in the services sector.

3.1 Framework conditions

One first indication for the readiness of services companies towards DSM acceptance and therefore, varying price signals, is the incorporation or use of flexible tariffs already today. Flexible tariffs are giving a price signal for customers to either reduce their electricity consumption at high price levels or to increase demand at low price levels. However, the majority of the companies participating in the survey are signing full supply contracts with only little or no transition to flexible or variable price signals (see Fig. 2). Depending on the market regulator, such dynamic price schemes are already today possible, but not yet fully made available to all clients.

The figure displays the share of respondents specifying their tariff structures within the individual countries. The majority of the companies are fully supplied with either fixed electricity prices or with day/night tariffs.

These results on tariff structures can be linked to the installed infrastructure (e.g., smart meters and the time resolutions of demand metering, see Fig. 3) and the availability and acceptance of such price schemes. In the UK, the share of companies metering at hourly or even sub-hourly level is highest with 38% and 15%, respectively. However, the number of companies with flexible tariff structures remains below 20%. In the other countries, only a minority is metering demand on hourly or sub-hourly levels (i.e., 5% and 20% in CH, 6% and 2% in PL and 12% and 2% in IT, resp.) and therefore, relevant demand data is missing which is needed to understand and potentially profit from DSM options.

Interestingly is the case of Italy. In Italy, a first smart meter rollout took place in the years between 2001 and 2006 with the installation of approx 30 million devices. However, looking at the feedback from the survey participants (see Fig. 3), the advantages of such devices for metering on hourly or sub-hourly levels are hardly

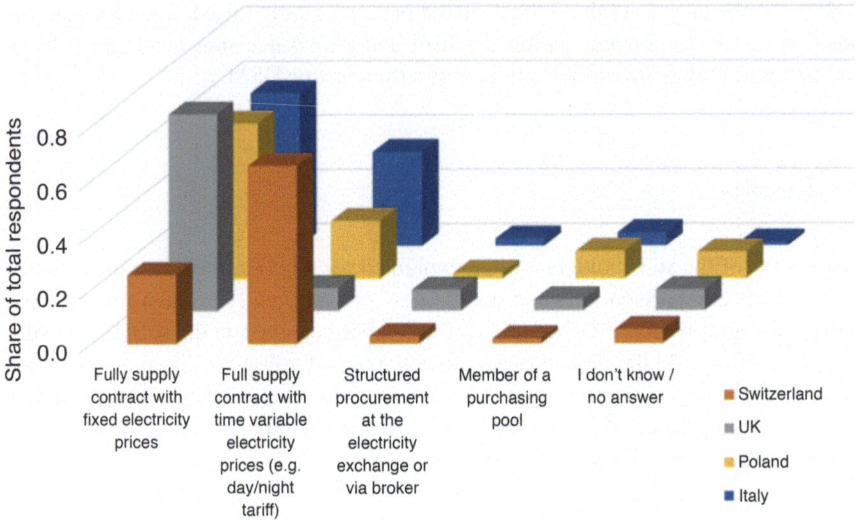

Fig. 2 Tariff structures

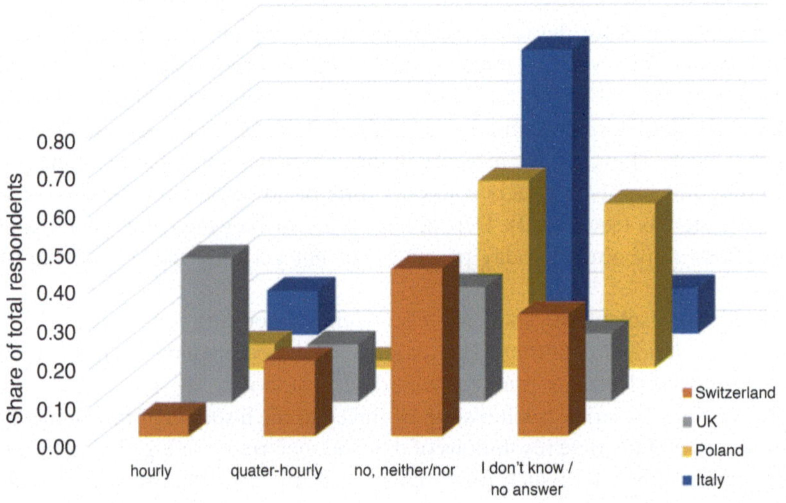

Fig. 3 Time resolution of metering

used. According to our knowledge, the utilities make this option to meter on hourly level only available for very large consumers which are reflected in the survey. However, from the answers we received, we can only speculate on the reasons why this service is only provided to large consumers as they could differ from more technical arguments (large data storage and handling needed, incl. technical facilities) to more financial aspects (the client has to pay for this additional service).

Therefore, utilities and electricity providers need to increase their efforts to activate the available infrastructure and to introduce supply contracts, including flexible price signals to allow the market to adapt to additional chances and risks of such price schemes.

The figure displays the share of respondents defining the time resolution of the metering system on site (either hourly, quarter-hourly or neither/nor).

Besides the questions on current state of electricity demand and its metering, the perception of energy efficiency and potential efficiency improvements are of relevance to understand the importance of energy costs for the current and future behaviour of the companies. In the survey, companies were asked if they had conducted an energy audit in the past 3 years (see Fig. 4). In CH and IT, 65–67% of the companies stated that they did not perform an energy audit in the recent past and in the UK (74%) and PL (80%) such shares were even higher. This non-performance of energy audits is disclosing a high potential for energy efficiency measures or higher shares of DSM systems to be implemented. These results were rather independent of the size of the electricity demand, showing shares between 48% (CH) up to 58% (UK, IT) of the companies with an electricity demand larger than 100 MWh per year which did not conduct an energy audit in the past 3 years. Only in Switzerland, the majority of companies with an electricity demand larger than 1 GWh per year did perform an energy audit in the recent years (i.e., 23 out of 35 companies with electricity demand >1 GWh did perform energy audit in CH).

The figure displays the share of respondents which have conducted an energy audit in the past 3 years.

Interestingly, only approx 23% of all energy audits conducted in the four countries were performed on mandatory base whereas 58% were voluntary energy audits (see Table 2). This is leaving room for further regulatory or incentive-based measures to increase the share of large consumers above 100 MWh energy demand to investigate their potential for efficiency improvements.

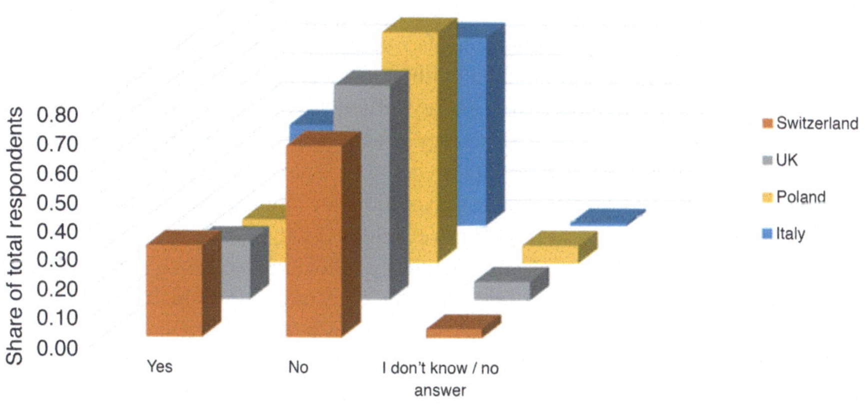

Fig. 4 External energy audit

Table 2 Overview on performed energy audits

	CH	UK	PL	IT
Voluntary energy consultation	61	25	21	65
Energy consultation within energy management system	12	9	4	11
Mandatory energy audit for large enterprises due to directive or legal requirement	15	16	12	27
I don't know/no answer	2	10	5	1
Total	90	60	42	104

Number of respondents reasoning why they have conducted an energy audit

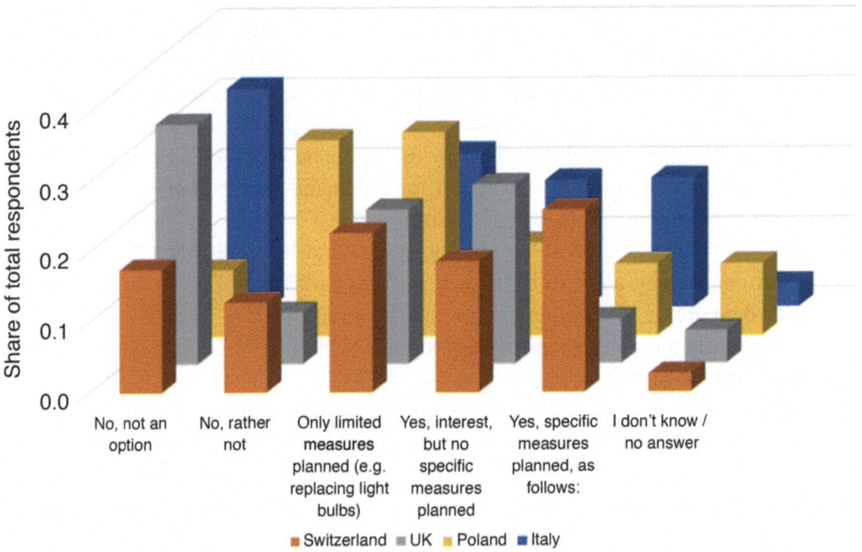

Fig. 5 Investing in energy efficiency

Given the expectation that larger consumers benefit the most from energy efficiency and/or a DSM implementation also under market conditions, additional efforts are needed that also companies with smaller energy demand look closer at the potential savings by improved energy efficiency or DSM implementation. However, the majority of large scale energy consumers (i.e., those most benefitting from its participation) are currently not involved in DSM.

In terms of planned and future energy efficiency improvements, the answers between countries vary only slightly. Between 32% (Switzerland) and 42% (UK) of the companies are less likely investing in energy efficiency measures in the coming 5 years (Answers "no" or "rather not", see Fig. 5). Companies which are planning to invest in energy efficiency, are more heterogeneously distributed and opt between "only limited measures" (e.g., replacing lights) and "specific measures" (up to 25% of the respondents in case of Switzerland). The performance of an energy audit strongly increases the probability for companies to invest in energy efficiency measures in the near future. Between 50 and 70% of the companies which did perform an energy audit indicate to invest in energy efficiency improvements in the near future (two categories "yes, interest"—with and without specific measures planned), whereas only 20–30% of the companies which did not conduct an energy audit show a positive interest to invest in energy efficiency measures in the short term.

Not surprisingly, a link between the companies having performed energy audits in the recent 3 years and those most likely planning energy efficiency measures within the next 5 years was observed.

The figure displays the shares of companies which are planning to invest in energy efficiency measures in the coming 5 years.

3.2 *Adoption of DSM*

Only few of Europe's service companies have so far adopted load management (see Fig. 6) with Switzerland showing the highest participation rate of 7%. Therefore, an insight on the potency of a DSM implementation of the non-performer can be provided. The general understanding and notion of DSM's and DR's versatility has certainly improved over the years [7]. However, almost independent of the regulatory framework in the countries investigated, limited numbers of service companies are participating in DSM as of today. Therefore, a supportive regulatory environment (e.g., CH) is not a sufficient condition to establish functioning DSM markets.

The figure displays the ratio of companies currently carrying out load management or not.

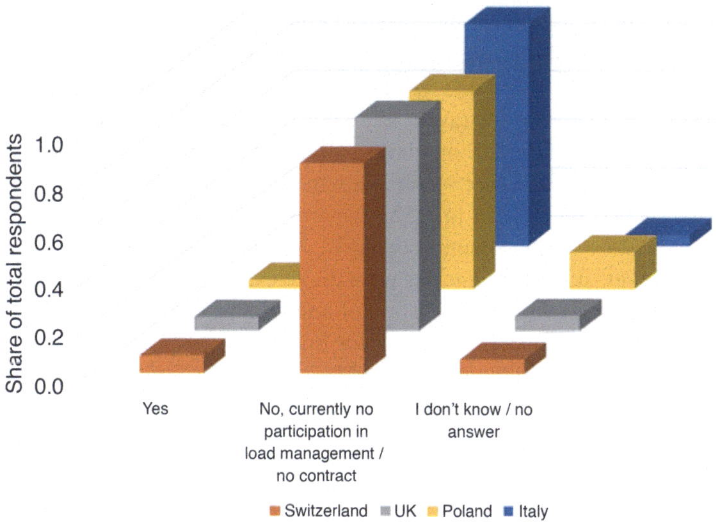

Fig. 6 Ratio of companies carrying out load management

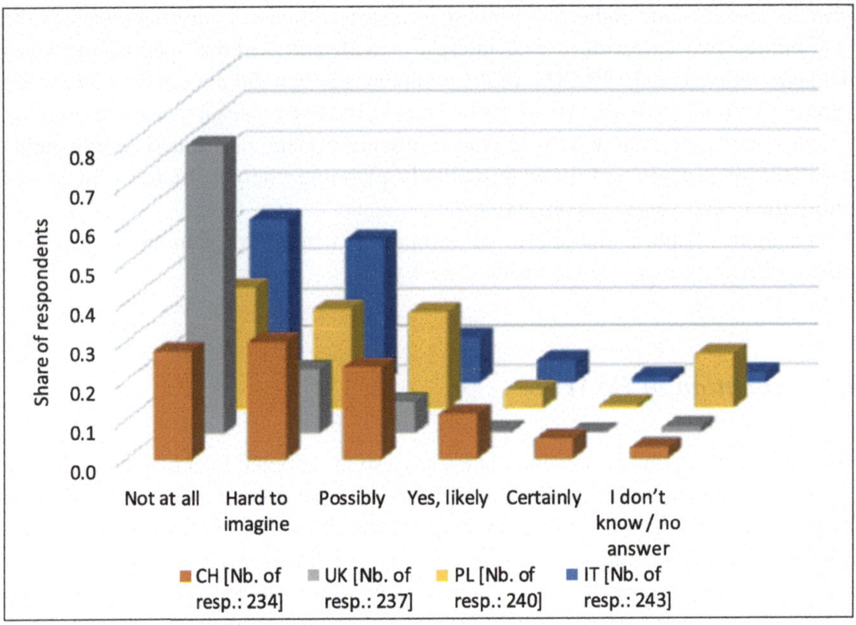

Fig. 7 Allowance of external control

As the verdict on the current participation in DSM is clear, the likelihood to participate in DSM in the future is of high interest. Therefore, participants of the survey were asked if they can imagine that external companies are allowed to control specific appliances on site to allow the service company to participate in DSM. The majority of the participants in all countries is stating that such interaction is hard to imagine or even not at all to imagine (see Fig. 7). Particularly the UK appears to disapprove of this possibility with 16% and 74%, resp., followed by Italian companies which disapprove by 37% and 42%, resp., although both countries possess at least one necessary condition for favourable DSM markets (favourable regulatory environment in the case of the UK and large roll out of smart meters in the case of Italy and partially UK). However, companies in Switzerland (23%) and Poland (25%) do show a certain interest in DSM (answer "possibly" in Fig. 7) and can partially imagine allowing for external controls. At the moment and based on the answers in the survey, it remains unclear what the relevant boundary conditions are for such differences and which combination of regulations, technical infrastructure and market conditions is needed to change for higher acceptance of external controls and therefore higher participation in DSM.

The figure displays the share of answers if the control of appliances by an external service provider can be imagined. The number of respondents is given in brackets.

Besides the named framework conditions, the risk perception as one of the expected drivers for DSM participation does not give a clear picture neither (see

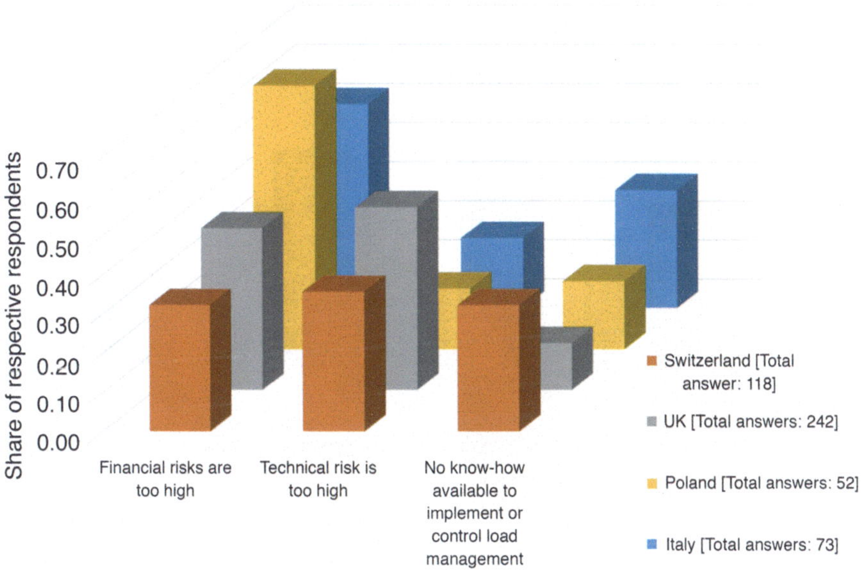

Fig. 8 Why DSM is not an option

Fig. 8), as the risk evaluation is not homogeneously distributed compared to the acceptance of external controls and the non-participation in DSM.

Companies which stated that they are not at all allowing external control of appliances or can hardly imagine to do so (see Fig. 7), were asked to state their main reasons for such selection (multiple answers allowed). While in PL and IT the financial risks of participating in DSM are perceived as too high (67% and 52% of the respective respondents), other factors such as limited in-house knowledge (32% in CH) or high technical risk perception (36% in CH and 47% in the UK) were the dominant reasons and need to be addressed by DSM providers to increase the number of potential DSM participants.

The figure displays the share of respondents giving reasons why DSM is not seen as option (in relation to the total number of answers stating that risks are too high, displayed in the legend of the graph).

Warren et al. stated in 2013 in their review of demand-side management policy in the UK [8] that it is clear that financial incentives and regulatory support are key determinants of the success of a DSM policy. In order to encourage participation in DSM, mostly monetary incentives are usually offered (ibid.). Therefore, participants were asked, which kind of incentives (monetary or others) they would expect if participating in DSM. Although for some countries, only few companies had a clear expectation, it is surprising to see that a majority of the respondents are willing to participate in the financial risk of DSM operations (see Fig. 9) by accepting a

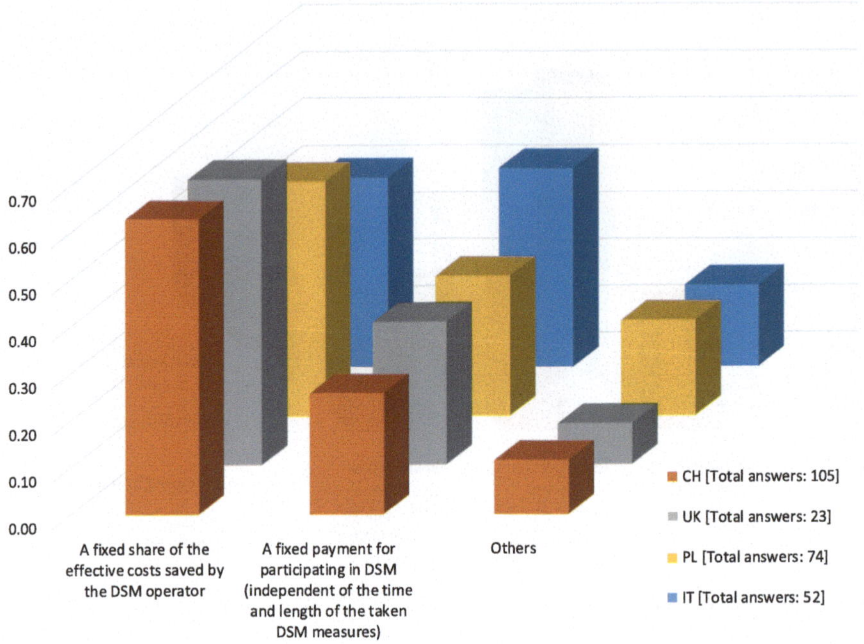

Fig. 9 Expected DSM incentives

fixed share of the cost saved by the DSM participation. By asking for a fixed share rather than a fixed fee for participation in DSM, the DSM operator or aggregator can transfer parts of the financial risk associated with the DSM offering to the client. Only in Italy, the interest in a fixed fee is higher as compared to receiving a fixed share of the revenues.

The figure displays the share of respondents stating their preferred revenue scheme for DSM (with either a fixed share of saved cost by DSM or a fixed payment for the participation of DSM. Number in brackets in the legend is the total number of respondents for this question).

Requesting a fixed share of the costs saved would leave the companies more freedom to handle appliance operation and might spark yet more interest in investing in energy conservation to save costs in the long term. A fixed payment for participating in DSM would not create any of such motivation, rather binding companies more to the specificities of DSM operations defined by the operator. Other companies don't agree with either and offered suggestions of their own but focusing on a general reduction in the energy bill.

In addition, compared to the effectively used remuneration schemes for DSM, the expectations are strongly deviating. Although the number of companies already today participating in DSM is small (see Fig. 6), some indications can be used for further discussion. Although on small numbers (CH = 21; UK = 17, PL = 9 and IT = 11), the fixed payment scheme seems to dominate besides other- more individual schemes. Only in Switzerland with 8 out of 21 respondents, a larger share of

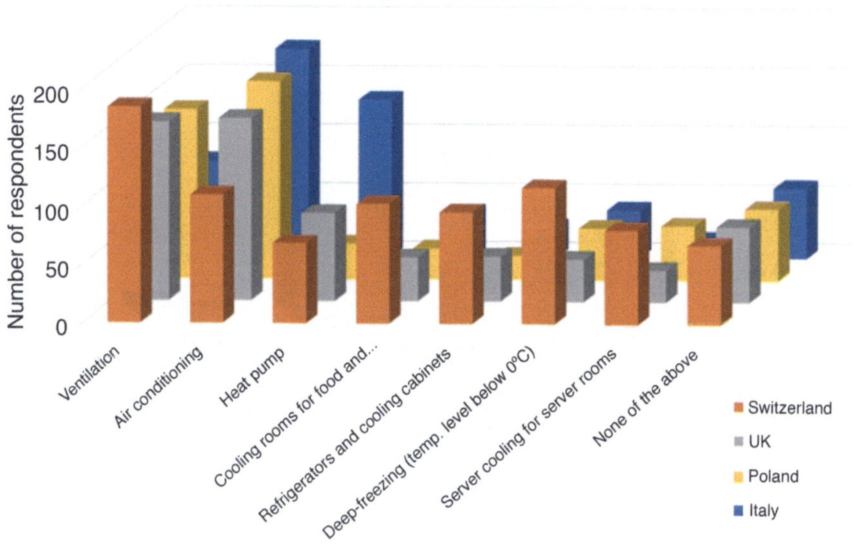

Fig. 10 Number of installed appliances

DSM participants is receiving a fixed share of the effective costs saved. In the UK and Poland, more companies receive a fixed fee for their participation in load management.

3.3 *Technical potentials*

Besides the expected revenues, the available appliances in the companies are of high interest. To estimate the future potentials of DSM, one needs to know which appliances are operated and under which conditions these appliances are operating. Therefore, respondents were asked, which appliances they operate (see Fig. 10) and if they would allow for external control.

The figure displays the effective numbers of different appliances installed in the four countries (multiple answers allowed).

Depending on the sub-sectors, some service specific appliances are installed but also other service-overlapping appliances are available and even in higher total numbers (see Fig. 10). All of the above-mentioned appliances are grid connected with an installed capacity of a few kW up to single units in the MW-range and therefore interesting for DSM operations in general terms. According to the survey, ventilation and air conditioning are the most commonly installed appliances. Depending on the specific conditions on site, these appliances offer flexible load shifting potentials for upward and downward flexibility [9]. However, as the hourly and seasonal load profile differ between the named technologies, not all systems can equally

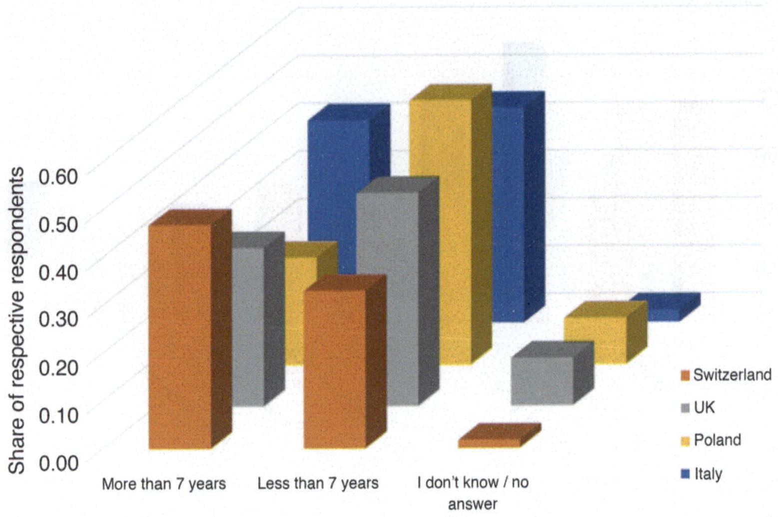

Fig. 11 Age distribution of potential DSM equipment on site

contribute to reduce electricity generation constraints based on high shares of intermittent renewable capacity.

Respondents were asked about the age of the installed equipment to better understand the potential development of increasing DSM options. It is expected that equipment which has not yet reached the end of life, will not be replaced in the short term. Moreover, companies are usually not ready to update new appliances shortly after installation, with the aim of keeping the system simple and additional configurations are supposedly time consuming and not always cheap to deal with. On the other hand, equipment which is already at the end of lifetime or beyond, offers the potential for fast introduction of DSM ready appliances. It is found that the age distribution varies strongly between countries (see Fig. 11) and appliances.

The figure displays the share for age distribution of installed equipment on site.

Looking at the country specific data on appliance related information from the survey, some additional indications relevant to accurately derive the DSM potential can be given:

1. In Italy, the high share of heat pumps is likely to consist of (reversible) air-air heat pumps [10]. Due to the limited shifting potential of air heating and cooling, the DSM potential is expected to be relatively small as compared to other countries with high shares of air/water and ground/water heat pumps (e.g. Switzerland).
2. In Poland, 65–70% of the installed equipment is in average less than 7 years old and therefore, replacement or updating with DSM ready control systems will be unlikely in the near future (add-on controls would only be expected with large installations or in case of overhaul/replacement).

3. In Italy, UK and Switzerland, the lifetime of the equipment is in average higher (in IT 41% of the equipment older than 7 years, in UK 48% older than 7 years and CH 59% older than 7 years) and therefore, an earlier exchange and upgrading of such systems can be expected, offering an installation of a DSM option.

Final survey questions were related to the decision processes within services companies. By asking about such processes and the purchasing structure of energy and the related topics within companies one tries to better understand the relevance of associated costs. It was found that the large majority of companies manage their energy administration at the highest decision level (exec. management or owner). Depending on the size of the company this is offering fast decision processes (e.g., small companies) or rather low relevance in case of large companies if no specific departments are responsible. Specialist with direct connection to energy demand and costs are usually more oriented towards new and innovative systems (i.e., DSM) and therefore more open to incorporating such flexibility options [11].

Given this overview and comparison of responses between countries, more country specific results can be found in the project deliverable [1]. Depending on the sub-sectors and distribution of respondents from the sub-sectors, the answers vary strongly between countries. However, such differences are the base for the adjustment of model input for other countries considered in the REFLEX model exercise.

4 Summary and Conclusions

With the survey-based analysis of service sector companies in four European countries, we cover a wide range of different DSM regulations and are able to depict the most relevant reactions of potential flexibility providers towards DSM participation. The survey results show that favourable regulatory conditions, such as in Switzerland and UK, are not sufficient to establish functioning DSM markets and attract companies' interest. As of today, particularly small demand units from services sector companies are hardly participating in DSM, although on aggregate level their DSM potential would be high. Lack of (reliable) information and of financial benefits as well as perceived risks seem to be relevant barriers against the adoption of more DSM option in the tertiary sector. Additionally, according to the survey results, there is a high potential for the introduction of energy efficiency measures and energy audits in the services sectors of the surveyed countries. The major shares of companies did not conduct energy audits in the last 3 years and they are not interested in implementing energy management systems. Meanwhile, 30–40% of the companies do not plan to invest in relevant energy efficiency improvements in the coming years. Altogether, the interest in optimizing and reducing electricity cost seems to be fairly low, although large-scale electricity consumers are reflected in this survey.

Regarding DSM, the current chosen tariff structures do not allow for load management to be implemented yet. This is partially expressing the low interest of companies to participate in DSM measures but is likely to be correlated with the low rate of high resolution electricity measurement needed for DSM. Except for Italy and partially the UK, where the installation rate of smart meters is high, most of the countries did not force a general roll out for smart meters yet. Therefore, utilities and the regulator need to adapt further to offer more flexible tariffs based on hourly or sub-hourly metering, investing the necessary resources in upgrading metering infrastructure (including Italy, where sub-hourly metering is scarce) and related data storage and analysis capacities. Additionally, the companies must adapt to more flexible tariff structures, which is risk related and could hence represent an obstacle to companies interested in DSM.

Few companies already today participate in DSM measures and even more, also small energy consumers use such options. Therefore, a potential for spillovers exist to introduce DSM in the services sector. However, the financial as well as technical risk implications play a crucial part in the implementation of load management. In each of the surveyed countries, these risks were stated most frequently, showing the need for appropriate risk mitigation options.

By showcasing existing examples of how flexibility providers can benefit from participating in DSM markets, we expect that risk perception of services companies can be altered. As the risk perception differs very much between countries, specific approaches (e.g., from aggregators) are needed to highlight potential gains. These gains can range between fixed recompenses up to variable, risk sharing, flexible payments. Besides such market-based transaction schemes, information regarding technical implementation, financial benefits as well as the functionality of DSM have to be provided to small and mid-sized companies as they lack relevant know-how. Specific information can be best provided by independent stakeholders such as governmental energy agencies or independent energy advisors. For small companies, such independent information is of high relevance to further promote DSM participation. Larger companies, with more standardized procedures regarding energy efficiency and energy demand (e.g., by implementing energy management systems) might adopt DSM models faster. Therefore, integrating DSM into energy management systems could support the further roll-out of flexibility options on the demand side. Energy advisors and auditors can help to spread the word of advantages and disadvantages of DSM systems.

Given the technical specifications of demand technologies, a large potential for DSM in Europe exists in services companies. As approx 80% of the companies investigated have at least one of the suitable DSM technologies installed or in use, aggregating such installations to reasonable DSM bids on the market seems to be achievable from a technical point of view. However, given the current status of regulations across Europe regarding aggregators and bid sizes, more efforts from policy makers are needed to reduce market barriers. Additionally, by integrating several technologies into one bid, transaction costs, as well as performance risks, can be mitigated. By aggregating different demand technologies into bids, one has to keep in mind the actual need of grid operators to manage demand by reducing or increasing load for specific time instances. As the described survey has only addressed demand in general terms, further information on the appliance specific hourly and seasonal load curves and their correlation with the expected generation load profile is needed to derive the full potential of the services sector DSM potential.

References

1. Reiter, U., Peter, R., Wohlfarth, K., Fermi, F., & Jakob, M. (2018). *Deliverable D2.2 – report on survey findings. Empirical study on DSM potentials and survey of mobility patterns in European countries*. Brussel: EU commission.
2. Behrangrad, M. (2015). A review of demand side management business models in th electricity market. *Renewable and Sustainable Energy Reviews, 47*, 271.
3. Reiter, U., Jakob, M., & Wohlfarth, K. (2018). *Policy brief–demand side management–Empirical data from the services sector.*
4. Arteconi, A., Hewitt, N., & Polonara, F. (2012). State of the art of thermal storage for damand-side management. *Applied Energy, 93*, 383.
5. European Energy Efficiency Directive. (2012).
6. SEDC. (2017). *Explicit demand response in Europe–mapping the markets 2017*. Brussels: SEDC.
7. Thies, F., Murray, G., Dong, J., & Bortolotti, M. (2014). *Mapping demand response in Europe today*. Brussels: SEDC.
8. Michaelis, J., Müller, T., Reiter, U., Fermi, F., Wyrwa, A., Chen, Y.-K., Zöphel, C., Kronthaler, N., & Elsland, R. (2017). *Comparison of the techno-economic characteristics of different flexibility options in the European energy system*. In: IEEE.
9. Klinke, S., Reiter, U., Farsi, M., & Jakob, M. (2017). *Contracting the gap–energy efficiency investments and transaction costs*. Bern: Bundesamt für Energie.
10. Warren, P. (2013). A review of demand-side management policy in the UK. *Renewable and Sustainable Energy Reviews, 29*, 941–951.
11. EurObserv'ER. (2013). Heat pumps barometer. EurObserv'ER.

Energy Consumption Monitoring and Building Performances in a Commercial Building: Case Study

Elena C. Tamaş (Papuc)

Abstract Our paper is based on a big size commercial building in Romania, in relation to the energy consumption monitoring, to the building energy performances and to the monitoring data analysis. It contains data which will allow us to compare an existing building with the retail industry norms in terms of the building energy consumption and energy costs. The building management system (BMS) is based on SAIA Burgess Controllers with IP interface, with input/output modules both analog and digital and remote I/O modules with S-NET interface collecting signals integrated in the entire building.

Based on the monitoring and data analysis, the paper's objectives are to present few action plans to implement them into this kind of buildings to achieve a better building energy efficiency. Additionally, to the main outcomes, the reduced energy costs obtained from the applied measures to improve the building energy efficiency building, there are mentioned other benefits arisen from the energy efficiency investments in building, like reducing energy use for space heating/cooling and water heating, improving interior thermal comfort, enhancing the property value, reducing the operational requirements and reducing the electrical energy for lighting, office machinery and other appliances.

Some recommendations are provided to control better and efficiently the building energy use and its consumption and reducing the avoidable waste while providing to the customers the services to achieve better interior thermal comfort conditions.

E. C. Tamaş (Papuc) (✉)
Faculty of Building Services, Technical University of Civil Engineering of Bucharest, Bucharest, Romania

P. Bertoldi (ed.), *Improving Energy Efficiency in Commercial Buildings and Smart Communities*, Springer Proceedings in Energy,
https://doi.org/10.1007/978-3-030-31459-0_2

1 Introduction

The Energy Performance of Buildings Directive (EPBD 2010/31/EU) [1] concerns the residential and tertiary sectors (offices, commercial buildings, public buildings, etc.). The energy performances of buildings prioritize three main aspects: (a) energy savings by improving mainly the buildings insulation; (b) increase energy efficiency through improved building installations; and (c) use renewable energy resources (e.g. solar energy and geothermal energy). Bloem [2] explained clearly three top consumption categories related to the building energy performances as follows:

- The building energy needs (the savings), which are directly linked to the indoor and outdoor climate conditions for working and living in buildings (by considering the comfort level of temperature, air quality and light, the temperature, solar radiation, wind);
- The building systems energy, by combining efficiently installations for heating, cooling, ventilation, hot water and electricity, is the relevant factor in the end-use energy consumption; the occupancy energy consumption (behavioral) covers all aspects considering the remaining use of energy function on how the occupant makes use of the building;
- Use renewable energy covered mainly by solar energy, bio-energy and to a smaller extent by geo- and aero-thermal energy will be a significant share in the final energy supply.

In Romania, all the above-mentioned are taken seriously into consideration when designing new buildings and when refurbishing old buildings. Our main laws on the energy efficiency and sustainable constructions are as follows:

- Romanian Law no. 121/2014 on energy efficiency [3] implementing Directive 2012/27/EU [4] on energy efficiency.
- Romanian Law no. 159/2013 [5], which transposes the provisions of Directive 2010/31/EC into national legislation, by amending Romanian Law no. 372/2005 [6] regarding energy performance of buildings (nZEB, EPCs, etc.).
- In line with the requirements of the Energy Efficiency Directive (2012/27/EU), the Romanian government approved the National Energy Efficiency Action Plan (NEEAP) [7] for 2014–2020, setting out the estimated energy consumption and planned energy efficiency measures to meet the 2020 energy efficiency targets.

Our analysis on the energy consumption monitoring and building energy performances was performed based on the 50,5 months monitoring starting with its first operational day following right after the building construction period. The monitored values were compared to the ones obtained for the designing phase and to the ones established in the norms. Some action plans for future improvements are proposed to obtain better building energy performances.

2 Building Assessment

2.1 Short Introduction of Building

The commercial building in question is located in Romania [8], and it was built on a 28.000 m^2 plot of land, having a 2.200 m^2 green area, a total gross building area of 114.000 m^2 (from which 45.000 m^2 above ground and 69.000 m^2 below ground), a total foot print area of 17.000 m^2, a total gross leasable area retail class A of 37.000 m^2, a total turnover rent area including storages and other leasable area of 41.000 m^2. The building has as urban indicators 65% P.O.T. and 3,5 C.U.T. The P.O.T. [9] is defined as the urban occupation percentage, and it is calculated as the percentage allowed to build the construction from the total land surface. The C.U.T. [9] is the land use coefficient, and it is defined as the built-up area of the building divided by the building land area.

The building has as main functions three basements (3B) from which two (3B, 2B) are mainly for parking and one (1B) is designated for retail, one ground floor (GF) for retail and coffee shops and two floors (1F and 2F) for retail, food, entertainment, offices and terrace. The second floor is used for offices, entertainment, food and a terrace for walking and other events. At each level there are technical rooms, corridors, toilets and staircases. The building has the following opening hours schedule: daily, from 8.00 a.m. to 10.00 p.m. for the whole building except for the terrace, the casino and the fitness areas. The fitness area is open daily from 6.00 a.m. to 11.00 p.m. The terrace is open daily from 8.00 a.m. to 12.00 p.m. and the casino is open daily from 8.00 a.m. to 2.00 a.m. (and in some occasions, at the tenant request is opened nonstop).

The building has 337.274 m^3 interior volume and 547.00 m^2 heated area for all the building levels. Regarding the thermal insulation performances, the building [8] has the following overall coefficients of thermal transmission: U = 0,30 W/m^{2}K for the wall cladding panels; U = 1,10 W/m^{2}K for the wall glazing; U = 0,25 W/m^{2}K for the roofs; U = 0,45 W/m^{2}K for the floor between the 2B (where there is parking) and the 1B (where there is retail) and U = 0,45 W/m^{2}K for the external doors.

2.2 Heating, Ventilation and Air Conditioning Systems Assessment

The general building design of the heating, ventilation and air conditioning (HVAC) systems [8] was made by defining the thermal modelling zones in the building levels. The general parameters to design the HVAC systems considered were: (a) air indoor set point temperatures; (b) the fresh air supplies; (c) the extract/recirculation rates; (d) the occupation levels; (e) the general lighting rates; (f) the internal heat gains; (g) the room destination; (h) the activity considered per room/area.

Our building was simulated over the winter/summer period using the typical weather city climate file containing hourly values for dry bulb temperature, wet bulb temperature, direct solar radiation, diffuse radiation, wind speed and wind direction. The heating energy demand for the building is ensured by two heating boilers having 1.950 kW each capacity and transported by 24 air handling units (AHU). Mainly for heating, there are air diffusers installed, but some areas (e.g. corridors, which are not used by the public) are heated with radiators. Simulations established that the air heating energy means around 80% of the total heat losses. The building cooling is made by two centrifugal chillers, each one having a total cooling capacity of 2.930 kW and the air distribution is performed by AHUs.

2.3 *Building Management System Assessment*

The building management system (BMS) is based on SAIA-Burgess controllers with IP interface with input/output modules both analog and digital and remote I/O modules with S-NET interface that collects signals integrated in the whole building. Each controller installed has a touch screen for local control of the system. Over the entire system there is a BMS server, which is a PC with a MOVICON 11 application, performing all functions required: view complete installation, view alarms, time programs, graphic visualization, history data base and multi-language functions. The software used is IEC 61131-3.

The system installed is modular and can be configured and extended according to the landlord's needs at any time by additional functionality, design elements and improvements. Other facilities provided by this system are as follows:

- Control unit and analysis of information monitor several addresses large enough to cover the needs of the beneficiary target object;
- Each addressing module has its own address and is controlled in the system, recorded and printed at any time in standby;
- Workstations for the detection and warning lines are changed from the control point, where disarmament to certain areas without affecting other work zones is possible;
- Surveillance of the sensors, printing and recording of work states are possible due to the ability to connect to displays with liquid crystal or via ethernet interface to a compatible PC and printer.

As an example, in the Fig. 1, we present some figures seen with the BMS to show, in this order, the counters for heating (column 2 and 3), cooling (column 4 and 5) and electricity (column 6 and 7) for each AHU registered in the building.

NO.	Meter ID	Meter index		Meter ID	Meter index		Meter ID	Meter index	
1	46104281	0043433 kwh	REPORT	47690346	0087764 kwh	REPORT	00015333	0068655 kwh	REPORT
2	46120040	0095955 kwh	REPORT	47690348	0129777 kwh	REPORT	00014598	0127162 kwh	REPORT
3	46120072	0047786 kwh	REPORT	47771222	0124351 kwh	REPORT	00014681	0135281 kwh	REPORT
4	46120078	0041801 kwh	REPORT	47697682	0107028 kwh	REPORT	00015765	0081692 kwh	REPORT
5	46120055	0137392 kwh	REPORT	47690347	0157181 kwh	REPORT	00015776	0123462 kwh	REPORT
6	46120045	0034456 kwh	REPORT	46079335	0223390 kwh	REPORT	00015304	0123763 kwh	REPORT
7	46120060	0089194 kwh	REPORT	47690349	0234436 kwh	REPORT	00014640	0237148 kwh	REPORT
8	46120056	0031607 kwh	REPORT	47690350	0230739 kwh	REPORT	00010594	0259343 kwh	REPORT
9	46120062	0039254 kwh	REPORT	47771217	0134361 kwh	REPORT	00014602	0127986 kwh	REPORT
10	46085818	0151672 kwh	REPORT	46079326	0003890 kwh	REPORT	00015760	0195920 kwh	REPORT
11	46085802	0086720 kwh	REPORT	46079334	0127730 kwh	REPORT	00017002	0371325 kwh	REPORT
12	47697688	0100568 kwh	REPORT	46079327	0035990 kwh	REPORT	00015317	0220655 kwh	REPORT
13	46104282	0057370 kwh	REPORT	47697685	0090439 kwh	REPORT	00014604	0216360 kwh	REPORT
14	46120036	0058313 kwh	REPORT	47697686	0129030 kwh	REPORT	00015272	0127231 kwh	REPORT
15	46085817	0163081 kwh	REPORT	46079325	0070740 kwh	REPORT	00014665	0361969 kwh	REPORT
16	46085816	0122728 kwh	REPORT	47690344	0017110 kwh	REPORT	00014671	0266800 kwh	REPORT
17	47694837	0209814 kwh	REPORT	47771233	0010825 kwh	REPORT	00014693	0186182 kwh	REPORT
18	46085795	0188339 kwh	REPORT	46079329	0111600 kwh	REPORT	00015327	0165627 kwh	REPORT
19	46104217	0000523 kwh	REPORT	47690345	0130631 kwh	REPORT	00014674	0087558 kwh	REPORT
20	46057236	0000176 kwh	REPORT	46027974	0000022 kwh	REPORT	00015273	0001323 kwh	REPORT
21	46104280	0044744 kwh	REPORT	47697687	0062700 kwh	REPORT	00015257	0066687 kwh	REPORT
22	46120054	0000025 kwh	REPORT	47694824	0000732 kwh	REPORT	00014636	0055825 kwh	REPORT
23	46142611	0013950 kwh	REPORT	46057246	0000166 kwh	REPORT	00014597	0007531 kwh	REPORT
24	46085803	0065194 kwh	REPORT	46079333	0186410 kwh	REPORT	00016748	0369383 kwh	REPORT

Fig. 1 Metering the heating, cooling and electricity for AHU

3 Energy Consumption Monitoring Analysis

The building monitoring was performed right after the opening day during the 4 operational years and 2,5 months, for the summer / winter period as follows: for the winter period (the heating time around 7,5 months) was between the 1st of January to the 15th of May and from the 15th of September to the 31st of December; and for the summer period (the cooling time, around 4,5 months) was between the 16th of April to the 15th of October.

The following information presenting the building monitoring was based on the BMS registrations. However, every month, together with the local maintenance teams, the local readings of all utilities supplies and energy meters were carried out to be sure that the counting system running with the BMS was working properly. On some occasions, there were situations when the reliable information was assured by the BMS and in others, because the local counters were not connected to the BMS, the information was based on the local readings. For these, in the next 2–3 months were made local and remote surveys and were compared to the previous months for the same period to update the system and functionality of the counters. For both heating and cooling services, the system was running with variable water flow. In this respect, the related pump group was equipped with a frequency convertor, which was installed into the control panel. The control was with variable pressure in the range of maximum and minimum pressure. In this way, each pump of the group became master, respectively backup pump.

According to our monitoring period, for a monthly average value, respectively for a yearly average value, as presented below in the Figs. 2, 3a, b, 4, 5a, b) the values for the energy consumption (in kWh) and utilities consumption (water in m^3, gas in m^3, electricity in kWh, lighting in kWh) for each summer–winter period were obtained.

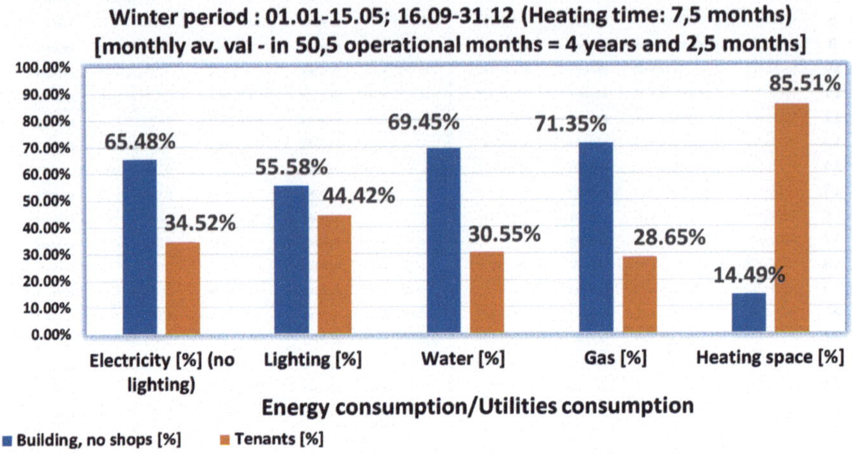

Fig. 2 Energy/utilities consumption during the heating monitoring period—monthly av. values (%)

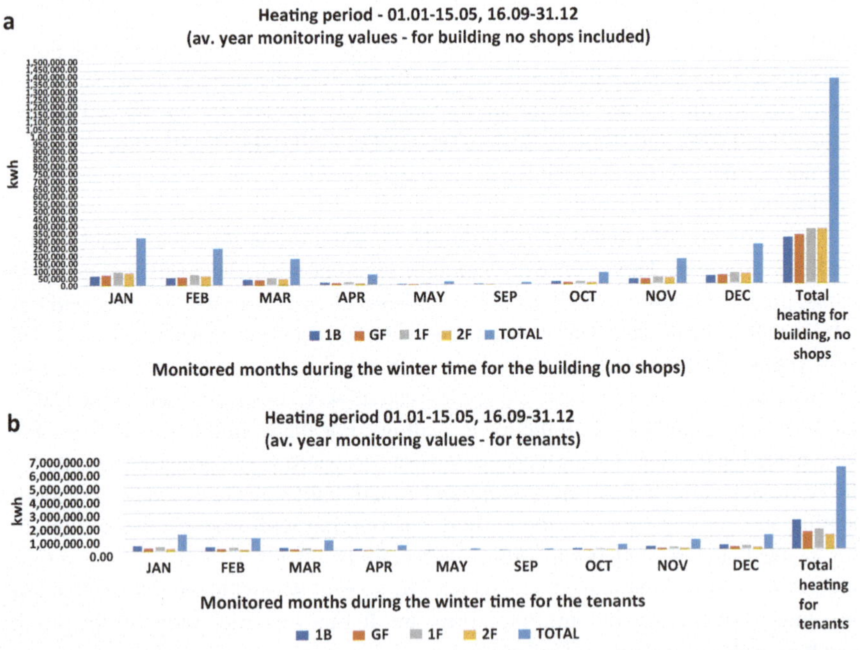

Fig. 3 (**a**) Heating period—monthly values for monitored building (with no shops) in an average year (%). (**b**) Heating period—monthly values for monitored tenants in an average year (%)

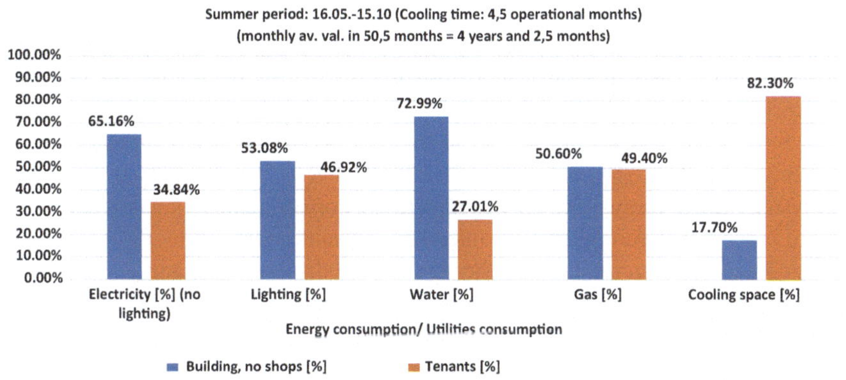

Fig. 4 Energy/utilities consumption during the cooling monitoring period—monthly av. values (%)

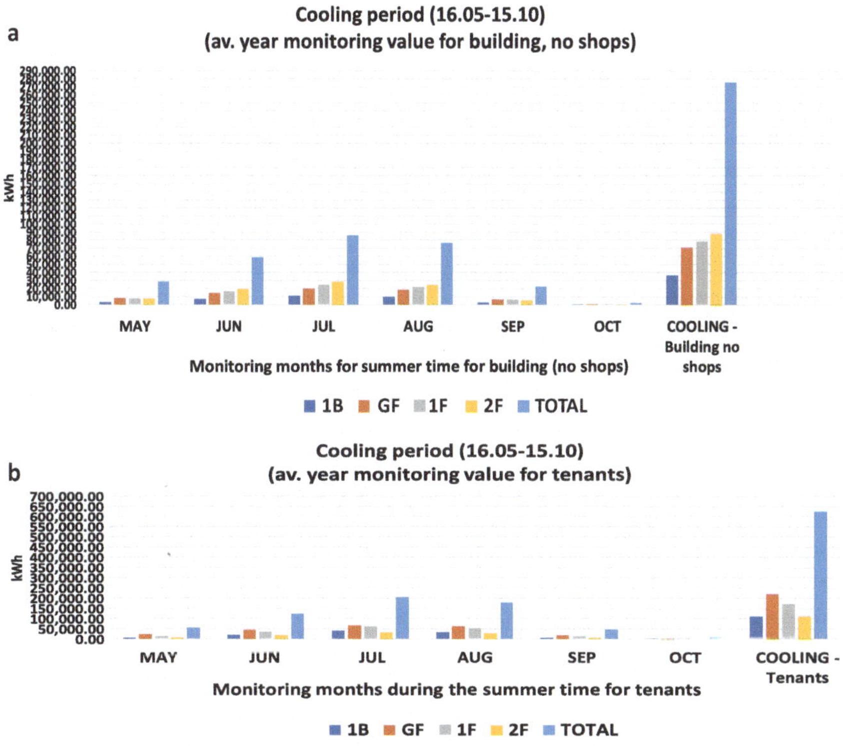

Fig. 5 (**a**) Cooling space monitoring period—monthly av. values for building, no shops (%). (**b**) cooling space monitoring period—monthly av. values for tenants (%)

The heating agent for the whole building is assured by the boilers with mixed burners: natural gas and diesel fuel. In case the natural gas distribution is damaged, or the natural gas supply is turned off for various reasons, the burners are switched for working with diesel fuel. Considering the temperature difference between the main boilers circuit (80 °C/60 °C) and the heating circuit which supplies the air handling units (AHU) heating coils (45 °C/25 °C) and the heat pumps (30 °C/25 °C), the system functions by using only a temperature correction injection into those two circuits. The pressure required for injection is ensured through by the circulation pumps fitted between the low loss header and the distribution manifold. The temperature correction circuit is dimensioned for a supply temperature of 80 °C and a return temperature of 25 °C. For the transition periods and the winter season, when the heat pumps work mainly in cooling mode, the dissipated heat by them is discharged into the AHU heating coils by using the low loss header between these 2 systems.

In the Fig. 2, there are presented in percentages for an average year, the energy consumption and utilities consumption per a monitored month during the heating period (winter time). The higher values obtained are for: the space heating consumed by the tenants (85,51%) from the total value of the space heating building; the gas consumption (71,35%) of the building with no shops; the water supply of the building with no shops (69,45%); the electricity (65,48%) and the lighting (55,58%) consumption of the building with no tenants included.

Figure 3a, b highlights the heating monthly values obtained during a winter in an average year for building, with no shops included (Fig. 3a) and for tenants (Fig. 3b). The peak value for the heating period is obtained during the month of January and the minimum is in September. For the building (with no shops included), 2F has the maximum value reached for heating, and for the tenants, this value is obtained on 1B.

The cooling agent for the AHU cooling coils is done by using the centrifugal water-cooled chillers, with a temperature of 6 °C/12 °C. The required condensation water for the chillers and for the heat pumps is supplied from 3 open cooling towers, each with 2 cells. The cooling towers have speed-controlled fans to control the output cooling capacity. The heat pumps are hydraulically separated by the cooling towers circuit by means of a plate heat exchanger, with primary/secondary temperatures 28°–33 °C/30°–35 °C. The expansion of the heating agent and maintaining the system pressure in the working limits is achieved by using closed membrane expansion vessels. The chillers supply chilled water at 6 °C for the supply and 12 °C for the return. The cooling towers operate with temperatures 28 °C/33 °C. The heating and cooling of the common circulation spaces on each floor are done and use compact AHU. These AHUs were dimensioned for functioning with 100% fresh air. The air is introduced using linear slot diffusers, each provided with plenum boxes and circular connections. The air is supplied around the mall areas along the perimeter at the limit between the tenants and circulation spaces. The exhaust air is taken from the space over the perforated false ceiling.

In the Fig. 4, there are presented in percentages for an average year, the energy consumption and utilities consumption per a monitored month during the cooling period (summer time). The higher values obtained are for: the space cooling consumed by the tenants (82,30%) from the total value of the cooling space building; the water consumption (72,99%) of the building with no shops; the electricity of the building with no shops (65,16%); the lighting (53,08%) and gas (50,60%) consumption of the building with no tenants included.

Figure 5a, b highlights the cooling monthly values obtained during a summer in an average year for building, with no shops included (Fig. 5a) and for tenants (Fig. 5b). The peak value for the cooling period is obtained during the month of July and the minimum is in October. For the building (with no shops included), 2F has the maximum value reached for heating, and for the tenants, this value is obtained on GF.

In a previous analysis [10] made for the commercial building no.1, which is our building presented in this paper, we have shown that, in an average year, considering the same monitored period and having 9/24/14/14 operational hours daily (as medium values), from the total building lighting consumption there are obtained: the highest value of 64,79% for the interior lighting building; 33,46% for the interior lighting parking; 1,28% for the building perimeter lighting and 0,47% obtained for the building exterior parking lighting.

4 Evaluation of the Building Energy Performances

In the Table 1, there is presented below a comparison of the building annual specific consumption for the building, as it was designed versus the building reference norms and their related energy class rating. It is shown that the building is rated with energy class A having an overall of 95,95% from 100% on the energy performances certificate.

Table 1 Comparison of the building annual specific consumption for design versus reference building norms

Building annual specific consumption (design versus reference building norms)			Energy class rating	
Installation type	Building design	Reference building	Building design	Reference building
Space heating	63,36 kWh/m^2 * year	31,18 kWh/m^2 * year	A	A
Water heating	21,36 kWh/m^2 * year	14,87 kWh/m^2 * year	B	A
Climatization	18,53 kWh/m^2 * year	27,60 kWh/m^2 * year	A	B
Mechanical ventilation	1,05 kWh/m^2 * year	1,05 kWh/m^2 * year	A	A
Lighting	45,02 kWh/m^2 * year	45,02 kWh/m^2 * year	B	B
Total building value	**149,32 kWh/m^2 * year**	**119,72 kWh/m^2 * year**	**A**	**A**
Equivalent specific annual emission indicator	**21,58 $kg_{CO_2}/m^2 * year$**	**17,32 $kg_{CO_2}/m^2 * year$**	**A**	**A**

Table 2 Comparison of the power energy consumption for the building heating obtained according design versus the monitored values during an average operational year (50.5 months monitoring)

Energy consumption for heating (kWh/year)	1B (kWh/year)	GF (kWh/year)	1F (kWh/year)	2F (kWh/year)	Building heating (kWh/year)
Design value	2.615.000	1.748.000	1.920.000	1.590.000	7.830.000
Monitoring value	2.450.000	1.650.000	1.810.000	1.380.000	7.290.000

Table 3 Comparison of the power energy consumption for the building cooling obtained according design versus the monitored values during an average operational year (50.5 months monitoring)

Power energy consumption for cooling (kWh/year)	1B (kWh/year)	GF (kWh/year)	1F (kWh/year)	2F (kWh/year)	Cooling Building (kWh/year)
Design value	150.400	297.000	248.000	207.000	902.400
Monitoring value	139.000	278.000	232.000	198.000	847.000

The overall total building value is for the building design as of 149,32 kWh/m^2 $*$ year versus 119,72 kWh/m^2 $*$ year for a reference building rated class A. Their equivalent specific annual emission indicators related to the overall building values are: 21,58 $kg_{CO_2}/m^2 * year$ for the building design versus 17,32 $kg_{CO_2}/m^2 * year$ for the reference building.

For the space heating type installation, which has a value of 63,36 kWh/m^2 $*$ year for the building design, we have an A energy class rate versus 31,18 kWh/m^2 $*$ year for the reference building corresponding to a B energy class rate. For the water heating type installation, which has a value of 21,36 kWh/m^2 $*$ year for the building design, we have a B energy class rate versus 14,87 kWh/m^2 $*$ year for the reference building corresponding to an A energy class rate.

The B energy class rate is obtained also for the lighting installation with a value of 45,02 kWh/m^2 $*$ year for both building design and reference building.

In the Table 2 [11] and Table 3 [8], we have a comparison made between the energy consumption shares of the building heating obtained according to the design versus the monitored values during an average operational year, having in calculation the entire monitored period of 4 years and 2,5 months. Because the building was executed following high standards in constructions and design, it was no surprise that the energy consumption values obtained for the heating and for the cooling during its operational years were very close to the reference building values.

5 Action Plans to Achieve a Better Energy-Efficient Building

For such energy-efficient building, even if it is well built and surveyed, some action plans are necessary to be taken to achieve an even better energy-efficient building and to maintain such level of energy efficiency in future.

The action plans to be performed, and proposed by the professionals in the field [12] and with which the authors are in line with, are as follows:

- Training both users and facilities teams to use, maintain properly all equipment, to monitor them closely, to intervene and to correct them as soon as possible;
- Hiring trained people with their relative skills for the specific services provider works;
- Monitoring and registered all energy consumption by meters linked to BMS and check them manually, periodically, with local readings;
- If possible to use it cabling for delivering the required power from inside/outside the building;
- Considering installation of solar PV panels (e.g., on the roof terrace) as an alternative to utility electricity supplies;
- Managing correctly and in due term the choice and monitoring the air quality filters to avoid damaging it equipment and cooling equipment for better performances, energy efficiency and reliability;
- Install humidity sensors and control them within a wider range of relative humidity to reduce humidification/dehumidification loads and the energy consumption;
- Considering using the free cooling technology;
- Replacing all non-performant lighting objects (incandescent/fluorescent) with the ones having LED light bulbs (LED technology);
- Reduce lighting consumption wherever building areas are not occupied and use motion sensors to activate the lights;
- For the cooling system, considering decreasing the condensing temperature and increasing the evaporating temperature to reduce the temperature variations to obtain a better efficiency and less work in cooling operation.

6 Conclusions

The design project building is very good and that is why it was rewarded with an A class rate for the energy performance certificate (95,95/100). This building heating is the most important energy consumption in the building and has a total consumption value of 7.290.000 kWh/year, of which 85,51% attributed to the tenants and 14,49% to the building with no shops included. The cooling building has a total consumption value of 847.00 kWh/year, of which 72,30% is attributed to the tenants, and 27,70% tor the building with no shops included.

It is important to acknowledge that by decreasing the design summer temperature by 2 °C, it is obtained a better interior building comfort and with a higher electrical cooling energy consumption. By retrofitting the incandescent exit signs with the LED lights, 95% energy consumption reduction in building lighting consumption can be achieved. By installing programmable sensors to control the HVAC equipment (and to schedule them and to set point reset capability) we can have significant energy savings for the commercial building analyzed (the exact values are still in working progress at this stage).

By installing adjustable/variable speed drives, the efficiency and providing a precise control over motor-driven loads (for the fans and for the pumps) can be increased. By installing an optimum start control on the air-conditioning system, daily, we also obtain energy savings. By installing plate heat exchangers on chillers (which it was done) at 12 °C below the chilled water return temperature, we can obtain a better thermal comfort and a free cooling available through these plate heat exchangers.

References

1. European Union: Directive 2010/31/EU of the European Parliament and of the Council on the Energy Performance of Buildings (recast). (2010). May 2010.
2. Bloem, J.J. *Energy and buildings. Renewable energies and energy efficiency*. Retrieved from https://ec.europa.eu/jrc/en/research-topic/energy-efficiency.
3. Romanian Parliament: *Law no. 121/2014 on energy efficiency*, Official Gazette Part I no. 574, August 2014.
4. European Union: Directive 2012/27/EU of the European Parliament and of the Council on energy efficiency, October 2012.
5. Romanian Parliament: *Law no. 159/2013 regarding the amending of Law no. 372/2005*, Official Gazette Part I no. 283, May 2013.
6. Romanian Parliament: *Law no. 372/2005 on energy performance of buildings*, Official Gazette Part I no. 662, May 2005.
7. Romanian Ministry of Energy: *National Energy Efficiency Action Plan (NEEAP)*, https://ec.europa.eu/energy/en/topics/energy-efficiency/energy-efficiency-directive/national-energy-efficiency-action-plans.
8. Tamaş, E.C., Ţârlea, G.M., Hera, D., & Flamaropol, G.C. *Energy consumption and data for the internal heat gains and opening hours at a commercial building*. Proc. of SGEM VIENNA GREEN 2017, International Scientific Conference on earth and Geo Sciences, Green Buildings Technologies and Materials vol. 4 (Hofburg, Vienna, Austria, 27–30 November 2017). ISBN 1314-2704, DOI https://doi.org/10.5593/SGEM 2017H, www.sgemviennagreen.org.
9. Romanian Parliament: *Law no. 242/2009 and its further updates on approving the Government Ordinance no. 27/2008 regarding the amendment of Law no. 350/2001 on landscaping and urban planning*, Official Gazette Part I no. 460, July 2009.
10. Tamaş, E.C., Ţârlea, G.M., Hera, D., & Flamaropol, G.C. *Status of the energy efficiency in retail industry*. Proc. of SGEM VIENNA GREEN 2017, International Scientific Conference on earth and Geo Sciences, Green Buildings Technologies and Materials vol. 4 (Hofburg, Vienna, Austria, 27–30 November 2017). ISBN 1314-2704, DOI https://doi.org/10.5593/SGEM 2017H, www.sgemviennagreen.org (Figure no. 3).
11. errata for table 1a from the paper written by Tamaş E.C., Ţârlea G.M., Hera D., & Flamaropol G.C.: *Energy consumption and data for the internal heat gains and opening hours at a commercial building*. Proc. of SGEM VIENNA GREEN 2017, International Scientific Conference on earth and Geo Sciences, Green Buildings Technologies and Materials vol. 4 (Hofburg, Vienna, Austria, 27–30 November 2017). ISBN 1314-2704, DOI https://doi.org/10.5593/SGEM 2017H, www.sgemviennagreen.org.
12. European Commission. 2018 *Best practices guidelines for the EU code of conduct on data centre. Version 9.1.0*, Joint Research Centre Technical Reports, www.ec.europa.eu/jrc.

National Energy Efficiency and Renewable Energy Action for Lebanon

Rami Fakhoury and Rani Al Achkar

Abstract The energy efficiency prominence in Lebanon has undoubtedly witnessed a significant change in the past decade. The National Energy Efficiency Action Plan—NEEAP 2016–2020 sets the road map for Lebanon towards achieving its energy efficiency (EE) and environmental objectives, and includes a number of EE initiatives targeting the different sectors of the Lebanese economy. Among these initiatives is the continuous support of the Central Bank of Lebanon (BdL) through a national financing mechanism (National Energy Efficiency and Renewable Energy Action—NEEREA), a program that has achieved a breakthrough in the sector since its launch in 2010, and has successfully catalyzed the EE market in an unprecedented way, thus putting Lebanon a step forward towards its target of achieving a demand growth control in order to save a minimum of 5% of the total demand in 2020 as per the Policy Paper for the Electricity Sector published by the Ministry of Energy and Water in 2010, and a 10% reduction in power demand through energy efficiency in 2030 compared to the demand under the business as usual scenario as per Lebanon's Intended Nationally Determined Contribution under the United Nations Framework Convention on Climate Change.

1 Introduction

On 21 June 2010, a national strategy targeting the electricity sector was adopted by the Lebanese Government through a policy paper. The Policy Paper included ten initiatives, among which three initiatives were dedicated to green technologies [2]. Among these is the need for national financing mechanisms. Therefore, on 25 November 2010, the Central Bank of Lebanon (BDL) issued Circular No. 236 set-

R. Fakhoury (✉) · R. Al Achkar
Department of Engineering and Planning, Lebanese Center for Energy Conservation (LCEC), Beirut, Lebanon
e-mail: rami.fakhoury@lcec.org.lb; rani.alachkar@lcec.org.lb

P. Bertoldi (ed.), *Improving Energy Efficiency in Commercial Buildings and Smart Communities*, Springer Proceedings in Energy,
https://doi.org/10.1007/978-3-030-31459-0_3

ting the conditions of application for green loans under the National Energy Efficiency and Renewable Energy Action (NEEREA), and thus the NEEREA financing mechanism was born.

The program supports the financing of all types of green projects, including energy efficiency (EE), renewable energy (RE), and green certified buildings.

It is undisputable that financing mechanisms are an extremely effective way to boost energy efficiency, renewable energy, and green buildings in a country. Fortunately, Lebanon is characterized by one of the best financing mechanisms in the region.

Regionally, similar programs do exist and possess the same objectives as NEEREA. In Jordan, the Jordan Renewable Energy & Energy Efficiency Fund (JREEEF) has established a financing mechanism for all RE and EE measures, covering the private and public sectors [3]. In Palestine, a revolving fund was established in 2012 to attract foreign funds. These funds are then invested in RE and EE measures to be implemented in public buildings [4].

NEEREA enables the private sector (individuals, small and medium enterprises (SMEs), or corporate bodies) to benefit from subsidized loans for any type of EE and/or RE projects. Loans are available to all sectors in the country, the subsidized and nonsubsidized (industrial, agricultural, residential, commercial, nonprofit organizations, etc.). NEEREA finances new environmentally friendly projects as well as the retrofitting of existing technologies with energy efficient products.

The characteristics of NEEREA are various; starting from the fact that loan amounts can be as low as 2000 US Dollars (USD) and as high as a ceiling of 20 million USD. The interest rate is low, typically from 0.3% to 1.075%, depending on the bank. Additionally, most Lebanese commercial banks are supporting the financing mechanism in order to enable the end users to have access to RE and EE technologies through the green loans. Till date, more than 17 Lebanese banks are involved in the NEEREA mechanism. Still, the most important aspect of the program is that it is a fully national mechanism, meaning that the incentives of NEEREA are based on incentives created and offered by BDL.

2 Foreign Incentives

During its start-up phase between 2011 and 2014, NEEREA benefitted from a grant offered by the European Union (EU) of 15 million Euros to SMEs. The EU has in fact supported NEEREA by contributing a grant over a share of the investment done by beneficiaries.

The grant is equivalent to 15% of the green loan amount for nonsubsidized sectors, and 5% for subsidized sectors, with a ceiling not exceeding 750,000 USD of grant money. Funds are allocated to the project after the Lebanese Center for Energy Conservation (LCEC) reviews and approves the technical study proposed. The grant money allocated would be only disbursed upon final execution and after technical validation by LCEC.

Till date, most of the EU grant money has been allocated by the BDL given the fast-growing use of NEEREA loans by beneficiaries. It is also important to note that the EU support to the BDL includes two major support tools: a small part of the EU grant is dedicated to finance the technical unit of the LCEC. Another part is dedicated to launch a nationwide marketing campaign to promote the use of green loans in the country. LCEC works as the technical arm of BDL in the review of NEEREA loan requests [5].

Additionally, NEEREA received a generous grant worth of 5 million Euros from the Italian Ministry of Environment, Land, and Sea (IMELS) in 2017 [6]. If 60% of the project cost consists of Italian products, then the beneficiary is eligible to a 10% grant on the project cost.

3 Types and Procedure

Since its start, there exists two types of NEEREA loans: If the facility is already operating, or existing, then the beneficiary can benefit from a repayment period of 10 years, including 2 years of grace period. Whereas, if the facility is under construction, or new, then the beneficiary will benefit from a repayment period of up to 10 years, with an additional grace period up to 4 years.

Additionally, green certified buildings can also benefit from subsidized loans. However, in terms of financing, contrary to other types of loans where the measure is fully financed, with green certified buildings, a certain percentage of the total project cost is financed; the higher the rating, the higher the percentage. The table below illustrates the different financing options for certified buildings (Table 1).

As for the procedure of obtaining the loan, the beneficiary can hire a consultant in order to prepare the required feasibility study. Once prepared, the file is submitted to the bank of his choice. In case the loan exceeds 20,000 USD, the file passes through the BDL before being reviewed by the LCEC technical team. If not, the file is directly reviewed by the LCEC before issuing the approval. Moreover, at the end of the project's implementation, a site visit is conducted by engineers of the LCEC in order to check the conformity of the project according to the design. On average, the whole process takes around 2–3 months, while the evaluation at LCEC takes 10–15 business days (Fig. 1).

Table 1 Distribution of financing under NEEREA [7]

Nature of project	Rating	Energy loan amount
New project	Not rated	Energy cost
	Certified	15% of total project value
	Silver	25% of total project value
	Gold	35% of total project value
	Platinum	45% of total project value
Existing project	Rated or not rated	Energy cost

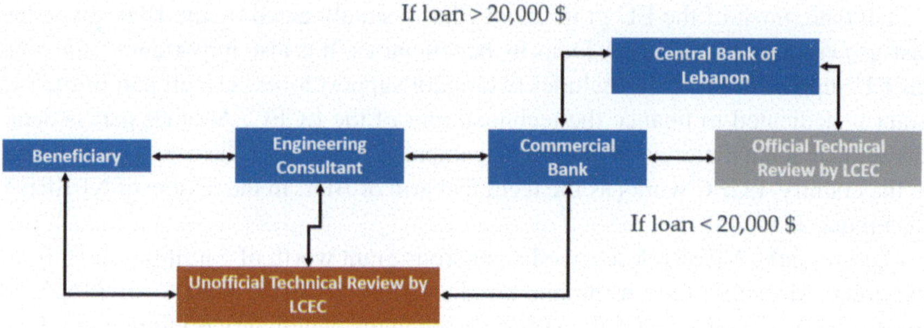

Fig. 1 Loan procedure [7]

4 Projects and Statistics

Although the list of financed measures is vast, the major and most requested technologies are few, as illustrated here below.

4.1 Light Emitting Diode (LED) Lighting

In terms of energy efficiency, efficient lighting systems took the biggest share in terms of number of projects. In fact, since the start of NEEREA, 115 projects have been implemented, taking into consideration new and refurbished projects, with more than 39,892 MWh of energy savings, equivalent to 4.3 million USD.

The pictures below illustrate a successful LED refurbishment project at the Hilton Metropolitan Palace in Beirut. The Hotel management successfully installed 24,320 LED indoor lamps, therefore reducing the peak load by 80%, and reducing the energy consumption by 2 GWh/year (Figs. 2 and 3) [7].

4.2 Building Envelope

Building envelope financing is another measure that has been growing in Lebanon, especially in the high mountain areas. The envelope improvement includes several items: double walls; insulation material for roof, walls and ground; stone cladding; double glazing. The number of projects till date is 92, with more than 18,000 MWh of energy savings, equivalent to 2 million USD (Fig. 4) [7].

Fig. 2 LED retrofitting project [7]

Fig. 3 LED retrofitting project [7]

4.3 Efficient Heating, Ventilation, and Air Conditioning (HVAC) Systems

Cooling and heating technologies are critical in a Middle Eastern country like Lebanon. Therefore, NEEREA finances all types of efficient cooling/heating technologies in order to reduce energy consumption as much as possible.

The number of projects till date is 60, with more than 16,000 MWh of energy savings, equivalent to 1.7 million USD [7].

Fig. 4 Photo showing double wall and insulation material during an LCEC site visit in Mount Lebanon [7]

4.4 Biomass Boilers

Biomass energy is also encouraged by the NEEREA financing mechanism. At present, more than 20 biomass projects have been financed, resulting in more than 481 MWh in energy savings, equivalent to 52 thousand USD [7].

4.5 Solar Water Heater

For solar water heating projects that exceed 5000 USD, the number of projects has reached 67, with more than 5718 MWh of energy savings, equivalent to 600,000 USD. (For individual solar water heating systems that cost below 5000 USD, the procedure is different and will not be discussed in this report [7].)

4.6 Solar Photovoltaic (PV)

As for the largest share of projects (in terms of RE), solar PV systems took the biggest share with more than 600 project till date, with an installed capacity reaching 25.13 MWp, and resulting in 38,326 MWh of energy savings, equivalent to 7.5 million USD.

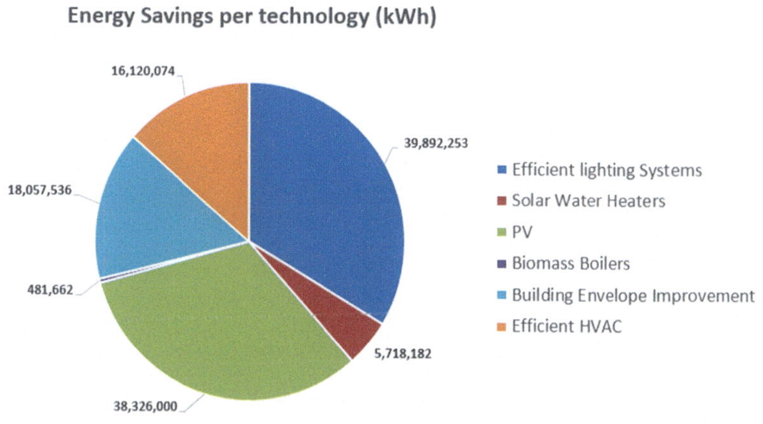

Fig. 5 Energy savings per technology [7]

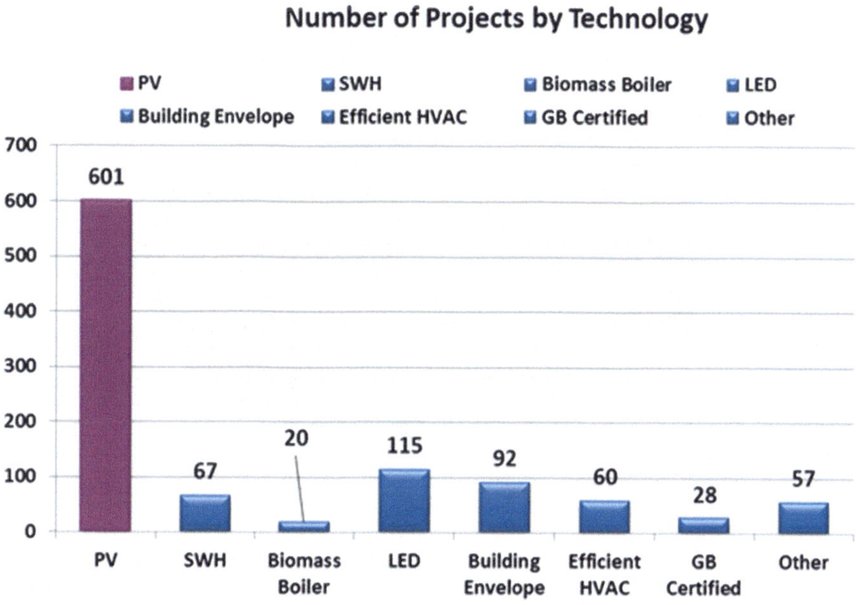

Fig. 6 Number of projects by technology [7]

Among these projects, 10.3 MWp were installed in the commercial sector, resulting in 16,123 MWh of energy savings, equivalent to 2.7 million USD [7].

The below figures illustrate the distribution of energy savings and of loan amounts across the different sectors and technologies (Figs. 5, 6, 7).

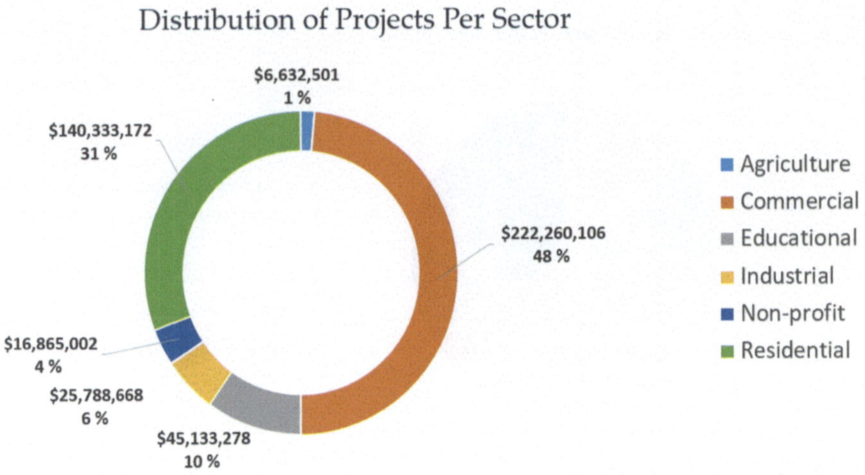

Fig. 7 Distribution of projects per sector [7]

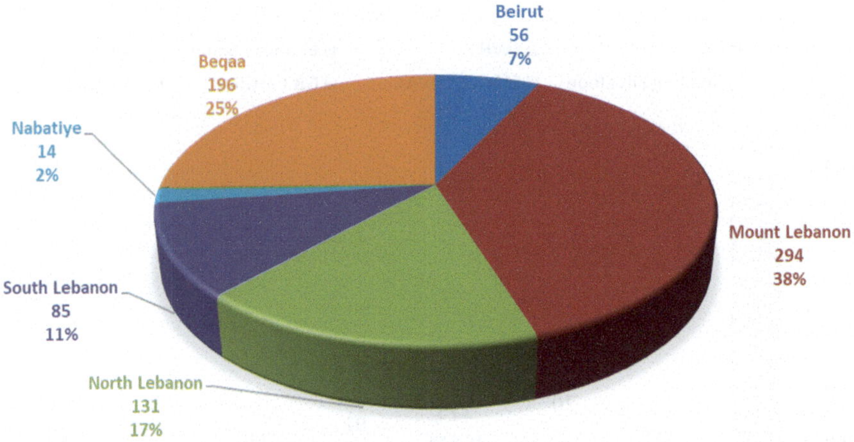

Fig. 8 Geographical distribution of projects [7]

Moreover, below is a graph showing the geographical distribution of the loans, followed by a map of Lebanon (Figs. 8 and 9).

Additionally, the growth in the number of projects from 2012 till 2017 of NEEREA projects is referred to in Fig. 10:

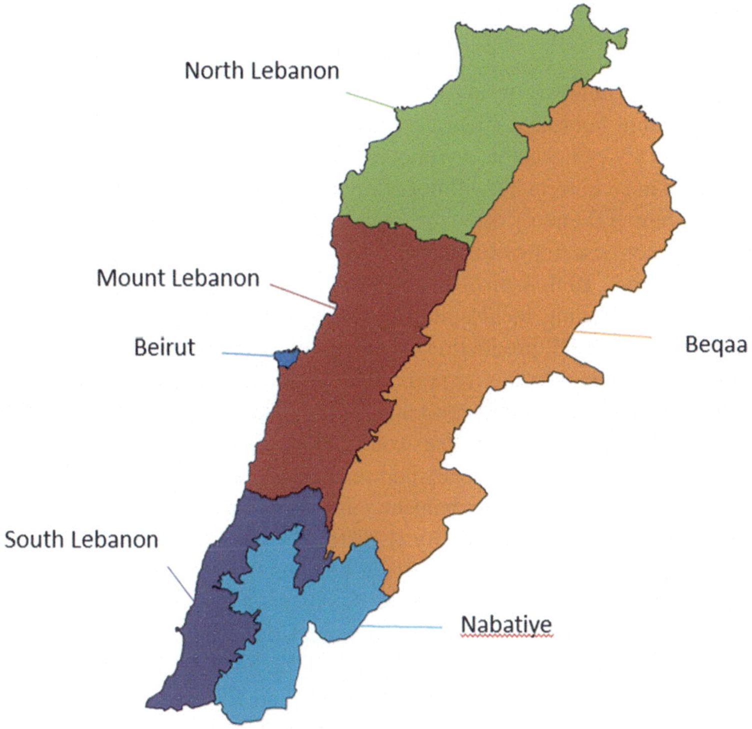

Fig. 9 Governorates of Lebanon [7]

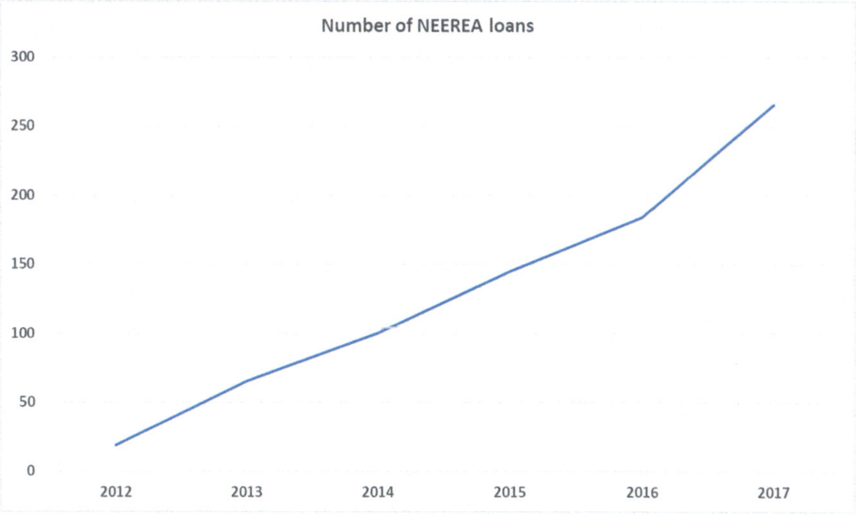

Fig. 10 Growth trend of NEEREA projects [7]

5 Next Steps

The LCEC is continuously trying to link energy efficiency to renewable energy. Therefore, starting February 2018, an energy audit has become mandatory to all NEEREA Solar PV projects that are above 60 kWp and to be implemented on existing facilities. This requirement will hopefully boost the EE market forward and will create more than 100 energy audits per year, thus benefitting all stakeholders and definitely the energy sector of Lebanon.

On another front, BDL aspires to extend the NEEREA financing mechanism to the public sector. Having local authorities benefit from subsidized green energy loans would boost the RE production significantly and reduce energy consumption throughout the country. In this regard, BDL issued the intermediate circular no. 399/2015 with an aim to support rural areas and villages [8]. Since then, LCEC has been setting the legal framework that would allow local authorities to leverage on this circular in their energy efficiency and renewable energy financing.

Also, another financing mechanism, the Lebanon Energy Efficiency and Renewable Energy Financing Facility (LEEREFF) has been built on the existing NEEREA financing mechanism. LEEREFF benefits from 80 million Euros funded by the European Investment Bank (EIB) and the Agence Française De Developpement (AFD). Similar to NEEREA, the funds are channeled to BDL and then to the private sector through the Lebanese commercial banks. The borrowers benefit from subsidized interest rates over 80% of the cost of RE and EE measures that are to be implemented, and 36%/28% of the project cost of Platinum/Gold green buildings, respectively [9].

Benefiting from the support of the BDL and the political commitment of the Ministry of Energy and Water, the LCEC has been continuously setting ambition targets and achieving remarkable milestones in improving EE, not just in commercial buildings, but in all sectors of the economy.

References

1. Bassil, G. (2010). *Policy paper for the electricity sector*. Retrieved June 2010, from http://databank.com.lb/docs/Policy%20paper%20for%20the%20electricity%20sector%202010.pdf.
2. Mortada, S. (2016). *The second national energy efficiency action plan for the Republic of Lebanon NEEAP 2016–2020*. Beirut: Ministry of Energy and Water/Lebanese Center for Energy Conservation (LCEC).
3. Ministry of Energy and Mineral Resources—Jordan. *Jordan Renewable Energy & Energy Efficiency Fund JREEEF*. Amman: Ministry of Energy and Mineral Resources - Jordan.
4. Myrsalieva, N., & Barghouth, A. (2015). *Arab Future Energy Index™ (AFEX) energy efficiency 2015*. Cairo: RCREEE.
5. Al Assad J. (2016). *The national renewable energy action plan for the Republic of Lebanon 2016–2020*. Retrieved November 2016, from http://lcec.org.lb/Content/uploads/LCECOther/161214021429307~NREAP_DEC14.pdf.

6. Italian Embassy in Beirut. (2017). *Italian participation at IBEF energy forum, 18–20 September 2017*. Beirut: Ambasciata d'Italia Beirut.
7. LCEC. (2018). www.lcec.org.lb. Retrieved 2018, from http://www.lcec.org.lb/en/NEEREA/StatisticsResults.
8. Association of Banks in Lebanon. (2015). *2015 annual report—Activities of the association of banks in Lebanon*. Beirut: Association of Banks in Lebanon.
9. European Investment Bank. (2017). *Technical assistance for the implementation of the Lebanon Energy Efficiency & Renewable Energy Finance Facility (LEEREFF)*. Luxembourg: European Investment Bank.

DGNB Framework for "Carbon-Neutral Buildings and Sites"

Anna Braune

Abstract The German Sustainable Building Council (DGNB) is the German and international knowledge platform for sustainable building and provides the world's most advanced sustainable building certification system. Its aim is the planning and assessment of sustainable buildings and districts. With more than 2800 pre-certified or certified projects worldwide and as a market leader in Germany, the DGNB can look back on more than 10 years of experience in fostering and certifying sustainable buildings and districts.

Since its very beginning, the ambitious assessments have always been based on the entire life cycle of a building, including both embodied and operational carbon emissions and applying to new buildings as well as to renovations of existing buildings. In order to support a significant contribution from the construction sector to limit global warming and in order to make the implementation of the Paris Agreement of 2015 measurable within certified projects, the DGNB has developed a framework for "carbon-neutral buildings and sites." The new framework consists of the three elements carbon accounting rules, carbon disclosure rules and carbon management rules.

The new framework offers a wide range of possible applications. It can, for instance, have a possible positive impact on DGNB certification outcomes, offer a reliable basis for decision-makers regarding the aspect of green financing, help to establish appropriate regulatory instruments or serve for educational purposes.

The framework has been published as a preview version in May 2018 and is applicable for a testing phase since July 2018.

A. Braune (✉)
Deutsche Gesellschaft für Nachhaltiges Bauen—DGNB e.V.
(German Sustainable Building Council), Stuttgart, Germany
e-mail: a.braune@dgnb.de

P. Bertoldi (ed.), *Improving Energy Efficiency in Commercial Buildings and Smart Communities*, Springer Proceedings in Energy,
https://doi.org/10.1007/978-3-030-31459-0_4

1 Introduction

1.1 The German Sustainable Building Council and the Paris Agreement

Climate protection is one of the essential topics to be considered regarding the certification of sustainable interiors, buildings and districts. Within the DGNB System, this is reflected in the strong focus on the life cycle assessment for new constructions, whereas for the sustainable operation of buildings a good carbon balance during operation is crucial. With the version 2018 of the DGNB System, the carbon neutral construction and operation of buildings have been set as target marks. Thereby, the DGNB translates into accountable and measurable actions the commitments of the Paris Agreement from 2015, which aims at mitigating climate change and decarbonizing the global economy. For the building sector in particular, this means achieving carbon neutral operations for all newly constructed buildings at the latest in 2030, and—more importantly—achieving a climate neutral building stock in 2050. By introducing the DGNB framework for carbon-neutral buildings and sites, the DGNB strives to make a significant contribution in order to achieve this target.

1.2 Motivation and Objectives of the Framework

The DGNB framework for carbon-neutral buildings and sites shall represent a translation of the commitments from the Paris Agreement into specific actions of the building industry and shall make those actions measurable on the basis of actual projects. The development of the framework was further strengthened by recent studies: The study "*Klimapfade für Deutschland*" describes a nearly zero-emission building stock as a possible scenario [1]. The study "*Szenarien für eine marktwirtschaftliche Klima- und Ressourcenschutzpolitik 2050 im Gebäudesektor*" suggests transition paths which make a massive reduction of carbon emissions actually appear achievable [2].

The framework shall provide a reliable basis for political and financial decision-makers as further described below. Unexploited potentials regarding the achievement of carbon neutrality shall be uncovered and persons involved in the construction process shall be motivated to realize an effective joint action. This shall contribute to massively accelerate the current renovation rate, to make use of all organizational measures within the facility management and to encourage building users to actively support climate protection through their actions.

1.3 Target Groups

The framework addresses several target groups: planners and building owners shall be supported in their wish to foster climate protection and given orientation when renovating buildings or planning to build new ones. Building operators shall receive support on the continuous improvement pathway of their buildings towards net zero carbon. Furthermore, funders, investors and banks shall be provided with a reliable basis for decisions on the climate protection-oriented financing of projects. Last but not least, policy makers shall be given a tool for actually implementing the building-related German climate protection goals and be encouraged to change the current energy focused laws.

1.4 Definitions Used Within the Framework

The framework uses and is based on the following two definitions:

The term "carbon-neutral buildings and sites" is used for buildings and sites whose CO_2 emissions are zero or less in accordance with the defined accounting method. These are buildings and sites with an annual CO_2 footprint of zero for building operation.

Buildings and sites with an annual CO_2 footprint that is permanently below a building-specific CO_2 limiting line (with annual maximum values) from now until 2050 and reaches zero by 2050 at the latest are designated by the addition of "carbon-neutral by 2050."[1]

2 Implementation

2.1 Application Possibilities

Six possible applications of the framework are highlighted below.

CO_2 Management and Benchmarking

The framework can serve as a basis for reviewing the energy planning of new construction and renovation projects in order to align it with the climate protection goals. Covering an extended accounting scope which includes user electricity that—for energy-optimized buildings—often surpasses the energy flows required for the energy systems within the building, the framework makes a significant step towards an actual emissions reduction.

[1] Initial applications will involve looking for a designation for buildings and sites that meet these conditions.

The calculation of the current and future CO_2 emissions[2] of a building according to the carbon accounting rules established in the framework allows a consistent benchmarking of the carbon intensity with other buildings. The building-specific emission limit values defined for each year from the current date to the target year 2050 deliver clear guidelines for the ongoing operation, thus providing consistence from the planning to the actual operation of a building.

Communication of the CO_2 Performance

By means of the consistent carbon accounting rules, the framework supports a clear and transparent carbon-related communication of both the planned and achieved contribution to climate protection. Within a market with a large variety of green labels, offering a recognized standard is crucial in order to earn the stakeholders' trust. Furthermore, clear rules regarding the reporting and the communication of the carbon performance are established within the part of the framework designated as carbon disclosure rules.

Links to the DGNB System

The framework can be applied without DGNB certification. The achievement of carbon neutrality for a building or site may in future be acknowledged within a DGNB certification. In order to incentivize the striving for more sustainable and even carbon-neutral buildings and sites, the version 2018 of the DGNB System already provides the possibility to earn different bonuses.

Influence on Supply Chains and Climate Protection-Oriented Procurement

The implementation of climate protection measures always requires a holistic consideration and evaluation. This means, that for instance the increase of the building's energy efficiency is only a partial aspect, while the procurement of a climate protection-oriented energy supplier is another important feature to be considered. The choice of the electricity provider, as well as the technology and the energy sources utilized by the energy supplier, will have a direct impact on the carbon intensity of the procured electrical energy. For the carbon accounting according to the framework, the use of CO_2 emission factors corresponding to the actual CO_2 value per energy unit (based on consistent determination methods) will be obligatory.

With regard to the selection of construction materials for renovation or new construction projects, the topic of climate change will be of increasing importance in the future. With the extended accounting scope including the embodied emissions by means of the life cycle assessment, carbon intensity will be a crucial criterion to be taken into account.

---

[2] For simplification purposes, the term "CO_2 emissions" is used in this document as a synonym for the sum of CO_2 equivalents (in accordance with international standards).

Most probably, carbon intensity will play an important part as future measuring unit for suppliers, especially in the field of public procurement since municipalities, communities and cities have already set highly ambitious climate protection targets which go far beyond the legal requirements.

Decision-Making Tool for Green Financing Opportunities

In order to foster the financing of climate protection-oriented projects, decision-makers willing to direct cash flows into future-proof buildings need a reliable basis for their investment decisions. The framework provides such a basis through consistent carbon accounting and carbon communication rules fostering transparency. The "climate protection roadmap" further described below helps to understand and evaluate the status quo of each individual building regarding its pathway towards carbon neutrality.

Basis for Appropriate Regulatory Instruments

Some elements of this framework have been integrated into a discussion proposal by the DGNB which has been published in February 2018 and submitted to political decision-makers. This DGNB Statement pleading for a "Building Emissions Law" valid until 2050 (GEG 2050) describes clearly the contents a law consistently focusing on climate protection should contain.

The proposal incorporates the following four core requirements [3][3]:

1. The target figure must be the CO_2 emissions and not the primary energy demand of our buildings.
2. The evaluation must be completed using absolute CO_2 emissions limit values and not by means of theoretical reference buildings.
3. If target values are not met, a CO_2 fee will apply.
4. All evaluations, specifications and control mechanisms must be based on data from actual consumption measurements.

In the course of the next years, the DGNB framework shall support the aspects required in the GEG 2050 and demonstrate their validity in order to foster the implementation of such legislative proposal.

2.2 *Scope of Application*

While for new constructions, climate protection-oriented measures can be implemented with a relatively low effort as they may be already integrated in the planning, it requires some more efforts to implement them for existing buildings. Therefore, newly constructed buildings certainly offer the highest potential for carbon neutrality

[3] Full discussion proposal available online: https://www.dgnb.de/en/news/statements/GEG_2050/.

but considering the current new construction rate of ~1%, the challenge lies in the implementation of these measures within existing buildings, which thus represent the main focus for the application of this framework.

Building Types

The framework is applicable to renovated and newly constructed buildings as well as to existing buildings. For renovated and newly constructed buildings, the assessment is conducted on behalf of planned values while for existing buildings actually measured values are to be used. For communication, the addition "planning values"/"planned" or "measured values"/"measured" must be used, respectively.

Assessment Boundary

The assessment boundary is extended: In addition to the energy demand for conditioning the building (building energy), the user electricity is also taken into consideration. The new assessment boundary is the property (site).

3 Contents of the Framework

3.1 Structure of the Framework

The framework consists of the following three main elements:

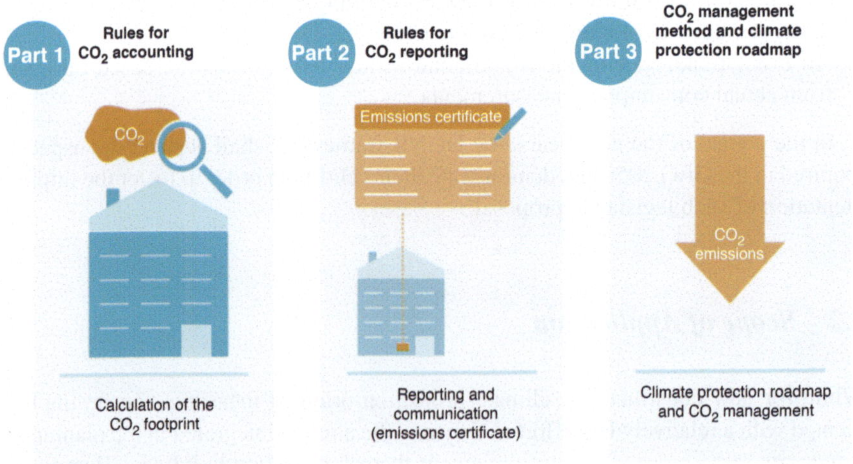

Three parts of the DGNB framework for carbon-neutral buildings and sites. © DGNB

These elements can be applied separately and reflect the interests and challenges of the different stakeholders. Part 1 addresses energy and specialist planners and describes how the carbon balance is calculated. Part 2 describes the minimum requirements regarding the reporting and the communication of the carbon performance which is done via an "Emissions certificate." In Part 3, a building-specific "climate protection roadmap" is set up providing information on the building's expected future carbon performance and its pathway towards climate neutrality. The essential elements of each part will be described subsequently. The entire framework is available on the DGNB website [4].

3.2 *Essential Elements of the Framework*

Carbon Accounting Rules

For the accounting of a building's carbon performance, two different accounting levels can be applied. Within "Accounting level 1," the balance framework only comprises the CO_2 emissions of the building operation.

Within "Accounting level 2," apart from the operational CO_2 emissions, the balance framework includes the embodied CO_2 emissions resulting from production, maintenance, dismantling or recycling processes.

Carbon Disclosure Rules

The "Emissions certificate" represents the essential element of the second part of the framework. It should be displayed directly at the building or site and be visible, understandable and present to everyone interested. The "Emissions certificate" shall provide transparency on actual results on the current carbon performance, on the planned carbon performance as well as on additional relevant CO_2 and energy parameters (per user or per floor area, final energy, share of renewable, etc.) Building users and visitors can thus always be informed about the current carbon performance and its expected development.

Carbon Management Rules

The essential element of the third part of the framework is the "climate protection roadmap" which is to be set up for each building individually. The roadmap demonstrates that the building/site will meet the 2050 target of zero CO_2 emissions in ongoing operation and that the total CO_2 emissions of the balance remain below the project-specific annual limits values.

In order to set up the roadmap, the building's current CO_2 emission level is to be measured serving as a building-specific starting point. As a next step, a straight line is drawn from this starting point to the zero-emission target in the year 2050, defining the building's individual CO_2 limit values for each year (see orange line in the illustration below).

For all buildings and sites with a climate protection roadmap according to the definition above, annual planning values will thus always remain below the building-specific emission limit values.

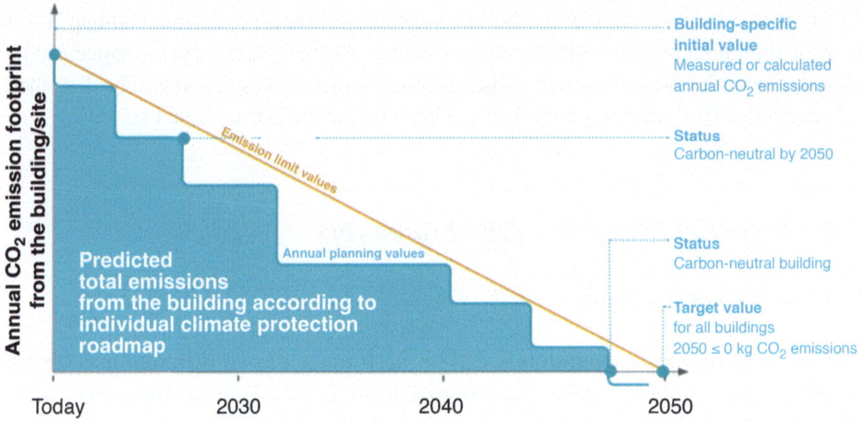

Climate protection roadmap as essential element of the DGNB framework for carbon-neutral buildings and sites. © DGNB

The "climate protection roadmap" must clearly state all planned steps and measures leading to carbon neutral ongoing operations. Furthermore, calculations according to the accounting rules defined in Part 1 of the framework must be presented for all these measures, reflecting the calculated carbon balance of the building/site for the remaining period until 2050. All measures planned and actually implementable for the building can and shall be included into the "climate protection roadmap" (demand reduction, improvements to the building envelope, installations, digitalization measures, improvement of energy production including future CO_2 factors, organizational measures, etc.).

References

1. Gerbert, P., Herhold, P., Burchardt, J., Schönberger, S., Rechenmacher, F., Kirchner, A., et al. *Klimapfade für Deutschland. The Boston Consulting Group GmbH*. Retrieved 2018, from https://bdi.eu/publikation/news/klimapfade-fuer-deutschland/.
2. Deutsche Energie-Agentur GmbH (dena). (Ed.). *Gebäudestudie: Szenarien für eine marktwirtschaftliche Klima- und Ressourcenschutzpolitik 2050 im Gebäudesektor*. Retrieved 2017, from https://shop.dena.de/sortiment/detail/produkt/szenarien-fuer-eine-marktwirtschaftliche-klima-und-ressourcenschutzpolitik-2050-im-gebaeudesektor/.
3. Deutsche Gesellschaft für Nachhaltiges Bauen—DGNB e.V. *DGNB statement—Proposal: The contents of a future German Building Energy Law in just three pages*. Retrieved 2018, from https://www.dgnb.de/en/news/statements/GEG_2050/.
4. Deutsche Gesellschaft für Nachhaltiges Bauen—DGNB e.V. *Framework for "carbon-neutral buildings and sites"*. Retrieved 2018, from https://www.dgnb.de/en/council/publications/reports/.

Economic Appraisal of Introducing Energy Efficiency in the Public Sector: Overview of Existing Economic Methods with Ex-post Application to Sustainable Energy Management Program in Croatia

Mia Dragović Matosović and Marko Matosović

Abstract Sustainable energy management system is an important prerequisite for making informed choices about which buildings to retrofit, choosing from the entire existing building stock. However, the system itself does not say anything about the economic potential of the retrofit, what economic methods to use for appraising buildings and what indicators to look at if the aim is the highest monetary savings. This paper demonstrates to local and national governments as owners of public buildings; (1) how they can use existing data from energy management information system (EMIS) to achieve highest monetary savings; and (2) it is economically justified to implement EMIS to benefit from the highest monetary savings when choosing buildings for retrofit.

The hypothesis was tested on a stock of 602 public buildings from Zagreb, Croatia. Through an Intelligent Energy Efficiency project ZagEE, Zagreb city administration chose 87 buildings for energy retrofit. The estimated economic savings were compared to a scenario where optimal 87 buildings would be chosen. The difference in savings justifies the additional cost of performing energy audits of the entire building stock. Although monetary savings are only one of the reasons for choosing the order of buildings to be retrofitted, the authors argue that additional savings can result in such greater savings that additional buildings can then be retrofitted, and other objectives achieved. The intention is to raise awareness of the savings potential that is exposed when investing into an efficient EMIS and properly using its results to decide which buildings to retrofit.

M. D. Matosović (✉)
Institute for European Energy and Climate Policy, Amsterdam, The Netherlands
e-mail: mia@ieecp.org

M. Matosović
Department for Energy System Planning, Energy Institute Hrvoje Požar, Zagreb, Croatia

P. Bertoldi (ed.), *Improving Energy Efficiency in Commercial Buildings and Smart Communities*, Springer Proceedings in Energy,
https://doi.org/10.1007/978-3-030-31459-0_5

1 Background

1.1 Economic Appraisal Methods for EE in Public Buildings

Overview of Existing Methods

Quantifying energy efficiency targets and achieved savings has become obligatory in European Union since the Energy Service Directive (ESD) stepped into force in 2009. Although it is mandatory to express a target, it is left to each member country to decide how to quantify the benefits [1]. Along with the ESD, the European Commission has published The Impact Assessment Guidelines, which attempt to help decide which policies to implement when it comes to energy efficiency. Having no widely accepted tests means that comparison between countries is hard and unfortunately, due to high costs of calculating EE program effectiveness, smaller countries or public administrations just take for granted benefits that occur in more developed countries or regions, replicating the models without knowing exact benefits those same programs will have in their economic surrounding.

Much has been written recently in the scientific community about the appropriateness of different ways to appraise energy efficiency measures, especially after the introduction of the Energy Efficiency Directive in 2009. In inspecting this subject pertaining to public buildings, two important, interrelated economic and technical questions come up:

- A question pertaining to energy: Which energy indicator to use for measuring efficiency?
- A question pertaining to the economy: Which appraisal methods to use?

A conference proceeding from a 2015 Nordic Conference on Construction Economics and Organization [2] tries to answer both questions by giving an overview of current economic efficiency practices for EE present in scientific literature. This research produced 18 articles, concluding that there are three most common methods for testing direct costs and benefits of projects: Multi-criteria analysis (MCA) or using certain indicators to compare projects, in this case, indicators for demonstrating energy savings; cost-benefit analysis (CBA) and cost-effectiveness analysis (CEA); and models measuring macroeconomic impacts.

The first only looks at certain factors on a project-by-project basis, and the review of articles showed that the payback period is the most commonly used indicator for energy savings, followed by annualized in-vestments costs and, equally represented, annual savings in energy costs [2]. The research further concludes that there are many shortcomings to this method, for instance, the payback period "is considered a method of analysis with serious limitations because it does not account for the time value of money, risk and other important considerations such as opportunity costs" ([2], p. 424).

CBA and CEA are used for a program or portfolio level, but still focusing mostly on direct benefits, while only macroeconomic methodologies measure the full effects EE improvements have on the society and consider a holistic approach of all costs and benefits encountered when implementing a comprehensive EE program. However, a macroeconomic model is usually costly and timely for owners of public buildings, and until comparable and monetizable indicators are developed and tested for measuring multiple benefits, cost-effectiveness analysis on a program level remains an effective tool for owners of a large building stock to compare and evaluate their choices.

Variety of Cost-Effectiveness Tests

Finally, there is no one way to use the cost-effectiveness analysis, rather many country-specific energy performance assessment tools exist which provide results including the countries' climate and building specificities; some of those are presented in A critical review of commercial and institutional buildings published in 2016 ([3], p. 1036)

The US has a much more consistent and older practice of testing cost-effectiveness of energy efficiency programs. Five cost-effectiveness tests were developed in California in 1974 when the California Energy Commission was established ([4], p. 20). Those tests are:

1. Participant cost test (PCT)
2. Program administrator cost test (PACT)
3. Ratepayer impact measure (RIM)
4. Total resource cost test (TRC)
5. Societal cost test (SCT)

Each test observes EE costs and benefits from a different stakeholder's perspective (PCT for the participant, PACT for the utility and RIM for the nonparticipant), while TRC and SCT test the overall program efficiency, including both indirect monetary (TRC) and non-monetized benefits (SCT). The tests have been used since the 1980s by all states as a principal approach to testing energy efficiency program overall effectiveness ([4], p. 20). As EPA states in its cost-effectiveness manual, as many tests as applicable should be conducted to properly choose between different program benefits and optimize the costs, the incentives and their distribution among stakeholders [5]. TRC is the most used test amongst the 50 US States, but it is again stressed that using more tests gives a more accurate overall cost-effectiveness overview.

Critique of Existing Models and Call for Additional Research

Through literature overview, especially in conducted real-life retrofit programs, three major oversights were recognized, the how, the what and the when of measurement uncertainties:

1. How to measure the savings or incomprehensiveness—most of the economic analysis are observing only a project instead of program level;
2. What to include in the measurements, or ambiguousness—disregarding indirect costs and benefits, due to an inexistent methodology for valuating and quantifying indirect benefits, especially macroeconomic ones (jobs creation, investments stimulation, energy import reduction, etc.). Also, often overhead and incremental administrative costs such as insufficient building technical data or missing legal documentation (e.g., cadaster irregularities) are disregarded;
3. When to measure, or overrating—overestimating savings through simulating energy costs instead of ex-post analysis of actual savings, and through exaggerating future energy price increase.

It is important to transparently account for all costs and benefits. Possible costs and benefits, as well as the angle from which they are viewed (beneficiary, utility, or local government level) greatly change the cost-effectiveness of any EE policy or program. This is sometimes easily used accordingly, to show unrealistically greater benefits for least costs, resulting in lower project bankability and financial institutions diffident to invest in energy efficiency and trust the calculated returns.

In 2015, the Home Performance Coalition from the Unites Stated has launched a program with a purpose to improve energy efficiency screening and decision making in the public interest [6]. In the program they call for a revision of standard testing practices of energy efficiency cost-effectiveness, concluding that current testing standards are not appropriate for new energy efficiency initiatives. The program continues to say that many details are left to interpretation, which then results in different applications of the tests which are then applied "in ways that do not accurately reflect the value of energy efficiency" [6].

Most recently, International Energy Agency lists 15 benefits, shown in Fig. 1 ([7], p. 28). Although energy efficiency has become an important topic, finding its way in all areas of energy strategies and development, its impacts are still many times being assumed rather than measured and even when the benefits are being quantified the focus is mostly on energy, CO_2 reduction and the direct financial savings that result from energy efficiency measures. These debates about dependable and unified measurements are visible today in the "green" scientific community in many fields in the EU, as is the case in the new Energy Efficiency Directive one of the main upgrades was agreeing on common standards for calculating baseline consumption and valid lifetimes of specific measures [8].

Next, another issue arises from using ex-ante predictions of future savings as a base for economic analysis. It is widely cited in scientific papers that there is "discrepancy between modeled and real building energy use" ([3] and [9], p. 1036). A solution to this problem could be developing and keeping an inventory of building stock and its energy consumption such as a Sustainable Energy Management Software developed in Croatia that will be described in the next section. Such an

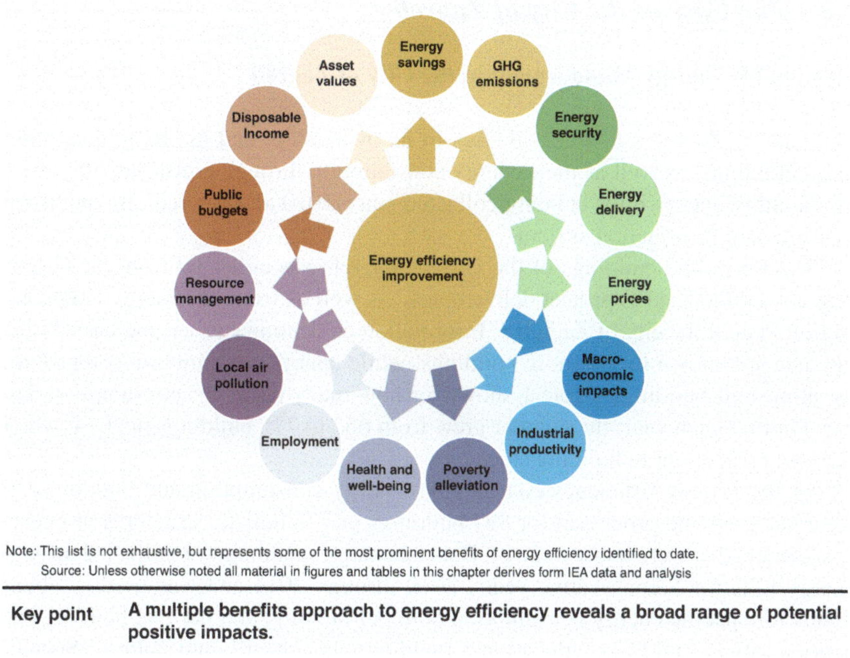

Fig. 1 Multiple benefits of energy efficiency improvements identified by International Energy agency [7]

inventory then provides a solid basis for monitoring actual changes in energy reduction after implementing EE measures ([10], p. 48).

A global sustainable policy has gained significance and the main challenge is no longer raising awareness, but rather how to develop it further, build its case and justify the cost-effectiveness of sustainable measures. This would be especially beneficial for politicians and regulators deciding on new energy pathways, financial institutions needing reliable calculations of return to minimize risk sensitivity, and would raise the market attractiveness of still somewhat new energy models such as energy performance contracting. Until all benefits that energy efficiency provides are not scientifically quantified, without favoritism or prejudice, there will not be enough incentive to turn to all those smart, existing solutions. Until then, authors opt for a cost-effectiveness analysis as a simple, yet accurate enough way to demonstrate actual monetary benefits.

1.2 *The Case of the City of Zagreb*

Sustainable Energy Management in the City of Zagreb

The City of Zagreb has 902 buildings in its ownership, and the basic data about these buildings, as well as their energy consumption through registering data from the monthly energy bills, is being collected since 2010 and entered into an energy management information system.

Data for public buildings of the City of Zagreb was collected from the City of Zagreb and the ZagEE project deliverables,[1] as well as from the Energy Efficiency Action Plan of the city of Zagreb.[2,3] Information on heating surface and basic building data are known for all those buildings, while energy consumption is registered for almost all buildings. Table 1 shows for how many buildings consumption data was familiar each year; the number grew from 652 to 776 buildings in 2014, which is about 80% of entire building stock.

For the Energy Efficiency Action plan, energy consumption and building data were gathered and processed for 897 buildings or 776 holistic structures or energy cost centers that were in use in 2014. Data on energy spending was collected through energy bills for 5 consecutive years, 2010 through 2014 and was used to obtain trends for specific energy consumption both per building and per five building categories: office buildings, educational buildings, healthcare, and culture. Specific consumption, as well as achieved EE savings, is usually given as final energy needs in kWh/m^2 of heated (useful) area ([11], p. 31). Also, energy usage was charted, dividing energy spent on biomass pellets, electricity, fuel oil, natural gas, centrally distributed steam, centrally distributed hot water, liquefied petroleum gas.

Table 1 Number of buildings in Zagreb, per usage type, for which the consumption data is available in the Zagreb Information system for energy management

	Year				
Type of buildings	2010	2011	2012	2013	2014
Culture	74	75	76	83	87
Education	375	378	380	376	395
Administration	78	96	112	151	159
Healthcare	123	125	126	126	132
Other	2	2	2	3	3
Total	652	676	606	739	776

[1] http://zagee.hr/?page_id=166.

[2] http://cei.hr/upload/2016/02/godisnji_plan_energetske_ucinkovitosti_grada_zagre_56bc950f80134.pdf.

[3] http://cei.hr/upload/2017/04/grad_zagreb_58e642252c88a.pdf.

ZagEE Project and the Retrofit of 87 Public Buildings

As the National Energy Efficiency Plan reports (2016), the project ZagEE—Zagreb Energy Efficient City was jointly prepared by the City Office for Energetics, Environmental Protection and Sustainable Development of the City of Zagreb, and the Agency, and they applied it to the IEE MLEI program (Intelligent Energy Europe—Mobilising Local Energy Investments). The scope of the project includes the energy retrofit of buildings owned by the City of Zagreb by applying energy efficiency measures and renewable energy sources for 87 public buildings; and the modernization of 3000 public lighting fixtures which will be replaced with LED lighting fixtures with a control and management system.

Project documentation estimates that the building sector in the Zagreb accounts for approximately 65% of total energy consumption of the city administration, and as such it carries the most potential for energy savings [12]. Project goals were, among others, assembly of quality technical documentation for the energy retrofit with a feasibility study for 87 buildings owned by the City of Zagreb, education of the project team in the city administration on the implementation of large energy retrofit projects and education of building managers on the efficient use of buildings. The results achieved through the project, especially the obtained technical documentation on the 87 chosen public buildings is the main dataset for performing the cost-effectiveness analysis in this research. Details about the 87 chosen buildings included physical characteristics of the building, as well as expected yearly savings in both Euro[4] and kWh/m^2. The methodology followed in this research and results are described in detail in the following section.

2 Methodology

2.1 Choosing Optimal Building Stock

Firstly, data used in the energy audits for 87 buildings and the data for those same buildings in the energy action plan (897 buildings) was compared based on m^2 and specific heating consumption (kWh/m^2). The inputs proved to be the same for both sets of data. The data used for further calculations included the following data on the 87 buildings[5]:

1. Heated area of each building in m^2;
2. Averaged total yearly energy consumption for heating in kWh;
3. Average yearly expenditure for heating from energy bills in Euro/a;

[4] The data was in Croatian currency, Kunas, but the calculations were performed in Euros, using an average exchange rate of 7.5 Kuna per Euro.

[5] The data is obtained from existing energy audits and the calculations were made according to the Croatian methodology for performing energy audits and energy certification of buildings.

4. Specific yearly consumption for heating in kWh/m^2 of heated area;
5. Expected yearly savings in kWh/m^2 and Euro/a after proposed interventions;
6. Expected yearly consumption for heating after proposed interventions, in kWh/m^2;
7. Total investment costs necessary for proposed interventions, in Euro.

From the former data, the following was calculated:

1. Specific yearly consumption for heating after proposed interventions, in kWh/m^2 of heated area;
2. Percentage of savings in specific yearly consumption per m^2 of heated area;
3. Price of an energy audit for each building, by maximum prices prescribed by the ministry of Physical planning and construction in 2010. The calculated prices for each building correspond to a realistic average of 0.67 $Euro/m^2$;
4. Trends in investment costs per type of building (low correlation), and per averages obtained from four quadrants (moderate positive correlation of 0.5). The latter is used (0.8 Euro/kWh saved for buildings in quadrant 1 and 1.47 Euro/kWh for buildings in quadrant IV);
5. Total lifetime savings, without discounting
6. Discounted energy net savings through a 25-year period (lifetime for service buildings was taken from the Croatian guidebook for measurement and verification of energy efficiency savings [13]), with discount rates 0–12% and energy price increase 0–12%;
7. Different scenarios of net savings without subsidies, with 20% subsidy and with a 40% subsidy on retrofit investment costs.

The latter data is all given and explained in the results section.

In all calculations, buildings were separated into building categories to establish any trends in consumption, possible savings or investment cost. Chosen categories, grouping buildings with similar building usage, are police stations, primary and secondary schools, kindergartens, retirement homes, health centers, and local boards (local government administration centers). All 87 buildings were given recommendations for a cost-effective retrofit, or approximately 50% reduction in specific final energy use per m^2 of heated area after implementing the proposed energy efficient measures. Proposed measures included:

1. building envelope insulation (separately the outside wall, connecting walls, floor and ceiling towards unheated area, and roof insulation were considered);
2. window replacement;
3. heating system change;
4. electric energy system interventions, and
5. Introducing renewable energy systems for own consumption.

The effect of measures was combined to get total savings, which resulted in a moderate retrofit of three to five interventions and average energy saving between 30% and 60%, as is defined in the 2011 Buildings Performance Institute Europe review of Europe's buildings ([14], p. 103).

2.2 Total Resource Cost Test and Anticipated Overall Savings

Finally, the calculated economic scenarios (net savings under different circumstances) were used to run a total resource cost test. Since this is a comprehensive test from the viewpoint of the utility or, in this case, the pro-gram administrator, all costs to the local government were included, to show a net benefit of implementing these measures. EPA defines the elements of the TRC test elements to be [5] (Table 2):

The TRC test was chosen since it views savings and costs form a program standpoint, but also since it does not include non-monetized (indirect, additional) benefits. A positive TRC result indicates that the program will produce a net reduction in energy costs in the utility service territory (or the owner of the savings) over the lifetime of the program. The importance of valuing additional benefits would be very useful to justify more of energy efficiency measures, but the challenge of monetizing indirect benefits from implementing energy measures is vast and goes beyond the scope of this research. Although bill savings are not regarded as a benefit in this test (since the test views the user perspective, the "user" here is the same City which is also the investor; the bill savings are included, under category "energy avoided costs." If the test was administered on households for which the City pays EE incentives, then the bill reduction would be disregarded. In this concrete research, the costs and benefits turned out to be the following:

The capacity-related costs avoided could further be expanded to included costs avoided for the energy utility if it is in the ownership of the program administrator. For example, if a comprehensive retrofit is done on most of the existing building stock, the city of Zagreb could save money on heat distribution costs, maintenance, and so on. However, in that case, the profit losses that will occur due to lower revenue collected from energy bills in those same buildings must be recorded as well. The resulting net benefit could be the difference in taxes avoided, or lower infrastructure and maintenance costs. Due to the complexity of the macroeconomic effects when viewing indirect costs and benefits, the non-energy benefits and indirect macroeconomic effects were not considered in this test. Instead, a straightforward approach was used to measure direct monetary costs and benefits.

Table 2 Listing TRC test costs and benefits, as advised by *EPA*

TRC test possible benefits	TRC test possible costs
Energy-related costs avoided by the utility (program administrator)	Program overhead costs
Capacity-related costs avoided by the utility, including generation, transmission, and distribution	Program installation costs
Additional resource savings (i.e., gas and utility, water if utility is electric)	Incremental measure costs (whether paid by the customer/building owner or utility)
Monetized environmental and non-energy benefits	
Applicable tax credits	

The U.S. Environmental Protection Agency [5]

The TRC test was administered for two different scenarios:

1. TRC test for the retrofit of 87 buildings nominated through ZagEE project, and the associated costs of energy audits for these buildings. This is the real-life state as was prepared in the City of Zagreb through ZagEE project; In this case, the costs and benefits were the following:

   (a) Costs:

   - Program overhead costs; a cost of current EE team in the City of Zagreb; Two employees, 6 months, 800 € net pay
   - Cost of energy audits (0.67 Euro/m^2 for detailed EPC) only for the 87 buildings (207,000 m^2)
   - Cost of main retrofit designs for ZagEE buildings; assumed at 1% of investment
   - Program installation costs; retrofitting 87 buildings taken from ZagEE technical documentation

   (b) Benefits:

   - Net energy savings for heating, (5% DR and 3% price growth) taken from 87 energy audits

2. TRC test for the retrofit of 87 optimal building, with costs for performing energy audits on the entire building stock. This is a more comprehensive approach and is suggested through this research.

   (a) Costs:

   - Program overhead costs; a cost of five employees, 6 months, 800 € net pay
   - Cost of energy audits (0.67 Euro/m^2 for detailed EPC) for entire building stock (1,429,454 m^2)
   - Cost of main retrofit designs for optimal buildings; assumed at 1% of investment
   - Program installation costs; retrofitting 87 buildings simulated from ZagEE data, as described in the results section

   (b) Benefits:

   - Net energy savings for heating, (5% DR and 3% price growth), simulated from ZagEE data, as described in the results section

Results of the two TRC tests are presented in the next section.

3 Results

Total resource cost test was administered on the two sets of 87 buildings; The ZagEE buildings and the optimal building stock identified through this research.

The results show that in the tested case and under described conditions, it is much more cost-effective to undertake a thorough building stock assessment before choosing which buildings to retrofit. Even with much higher costs of inspecting entire building stock of over 600 buildings and requiring 2.5 times more man days for administrative costs, such as collecting the data on building stock, it is quite cost-effective to undertake this systematic approach to viewing the building stock. Even with a large difference in overhead costs, the optimal chosen building stock results in a 43% greater net cost-benefit than when not all buildings are inspected before determining which ones to retrofit.

In this concrete example, many conditions were considered, as different discount rates, predicted price increase and different co-financing scenarios. The savings and the consumption were taken as-is from the provided energy audit data. This gave different outputs, or different absolute amount in net benefit difference between the first (ZagEE) and proposed (Optimal) building retrofit. If a realistic but conservative scenario is regarded (no co-financing, 2% price increase and a 5% discount rate), the result is the following:

ZagEE buildings return a 1,669,394 Euro net cost benefit, and Optimal building retrofit results in 4,317,601 Euro net cost benefit, while the costs are only 60% higher in the Optimal scenario which includes energy audits for the entire building stock as well as more time and resources spent on the energy efficiency team. This means that the proposed, comprehensive option of inspecting entire building stock before choosing which buildings to retrofit yields in this case with a 2.67 million Euro higher net benefit. This amount of monetary savings would be sufficient for:

1. Roughly, covering costs of an EE team of ten people for 10 years;
2. Performing detailed audits for approximately 2000 more buildings (e.g., schools 2000 m^2 in size);
3. Retrofit at least 11 more buildings to a modest retrofit level (average retrofit, in this case, was 240,000 Euro).

However, this approach does not account for different legal or organizational issues that might arise and influence the choice of buildings that will undergo energy efficiency retrofit.

These additional savings can also be used to develop and implement a continuous energy monitoring system, such as the Energy management information system mentioned in the case of the City of Zagreb.

4 Conclusion

A vast potential for energy and money savings that public buildings carry, is a well-tested and proven fact. However, hurdles still exist on the way to an efficient retrofit of public buildings. Even when such measures are undertaken, they tend to be unsystematic and the real cost-effectiveness is not calculated, nor are different scenarios compared to achieve the greatest savings. Both the results of this research, as well as the experience from practice have shown that it pays off to develop a systematic approach and first collect data on the entire building stock of one local administration body (e.g., city, municipality), before any further decisions about retrofit are made. Only after a rational indication of which buildings prove the most efficient for energy retrofit, can the most cost-effective decision to invest in retrofit be made. In this way, the most energy is saved, and this difference most of the time usually compensates for the extra costs needed upfront for a comprehensive building stock assessment.

This research and the example of ZagEE buildings have demonstrated that it is economically justified to invest in a comprehensive, systematic approach to listing and overviewing the entire building stock, along with its energy consumption. In this case, perhaps the reason for choosing exactly those 78 buildings in ZagEE project could have been organizational or administrative (ready technical documentation, existing interest of the building occupants, choosing evenly over City neighborhoods, etc.). The intent of this research is not to conclude that political or policy reasons are to be disregarded, but rather to point out that in the energy-economic sense, choosing optimal building stock can produce such greater savings which can then account for much higher costs of introducing a comprehensive energy management for entire buildings stock.

Costs and benefits of introducing energy efficiency into public buildings were described, with preference given to the cost-effectiveness analysis and showing different economic tests that can be best applied to assess the cost-effectiveness of such programs. A total resource cost test was performed on a set of buildings optimally chosen following the described method and the TRC test showed a 42% greater total net lifetime savings compared to buildings not chosen following the proposed methodology.

Once a proper economic method is applied to direct costs and benefits, there can be a clear overview of the net benefit of certain energy efficiency measures. However, a proper public body analysis should not only go beyond the project level to look at a program level, as shown here but instead, EE policies should be regarded on the portfolio level, with both direct and indirect costs and benefits considered. Only such measurement shows how much the beneficiary (the public sector, the government, or the community) needs to value the indirect benefits for the measures to be economically justified.

Proposed further research could be to develop methods to include those identified omitted indirect costs and benefits into a more holistic economic test for energy efficiency public EE portfolio appraisal. Researching ways to quantify additional benefits (include more of the indirect benefits, and their relevant costs) achieved in public sector energy efficiency programs would enable more programs to appear profitable. Such a thorough insight in all possible benefits, as well as a tested method of quantifying the results, could improve the financial attractiveness of energy efficiency programs; it would influence decision makers to focus more on energy efficiency measures, especially capitally extensive ones, with longer lifetimes. Raising awareness on the indirect, social prospects of energy efficiency would make both decision makers and financial institutions more lenient towards supporting more of energy efficient programs, both in public buildings, but across all other sectors as well.

References

1. European Parliament. (2012). Directive 2012/27/EU of the European Parliament and of the Council of 25 October 2012 on energy efficiency. *Official Journal of the European Union Directive*, (October), 1–56. https://doi.org/10.3000/19770677.L_2012.315.eng
2. Tuominen, P., Reda, F., Dawoud, W., Elboshy, B., Elshafei, G., & Negm, A. (2015). Economic appraisal of energy efficiency in buildings using cost-effectiveness assessment. Procedia Economics and Finance, 21, 422–430, https://doi.org/10.1016/S2212-5671(15)00195-1
3. Ruparathna, R., Hewage, K., & Sadiq, R. (2016). Improving the energy efficiency of the existing building stock: A critical review of commercial and institutional buildings. *Renewable and Sustainable Energy Reviews, 53*, 1032–1045. https://doi.org/10.1016/j.rser.2015.09.084.
4. U.S. Environmental Protection Agency. (2008). *Understanding cost-effectiveness of energy efficiency programs: Best practices, technical methods, and emerging issues for policy-makers. A resource of the national action plan for energy efficiency*. Washington DC: Office of Air and Radiation Climate Protection Partnerships Division. Retrieved November 12, 2015.
5. National Action Plan for Energy Efficiency. (2008). *Understanding cost-effectiveness of energy efficiency programs: Best practices, technical methods, and emerging issues for policy makers*. Energy and Environmental Economics, Inc. and Regulatory Assistance Project. (November), 96. Retrieved from http://www.epa.gov/cleanenergy/energy-programs/suca/resources.html.
6. Home Performance Coalition. (2014). *The national efficiency screening project mission statement*. Retrieved from Home Performance Coalition. http://www.homeperformance.org/policy-research/advocacy/national-efficiency-screening-project.
7. OECD/IEA. (2014). *Capturing the multiple benefits of energy efficiency*. Paris: IEA Publishing.
8. European Commission, DG ENER. (2012). *Energy efficiency directive (2012/27/EU). Energy efficiency directive (2012/27/EU) of the European Parliament and of the Council (L 312/1), 11,15,16*. Brussels, Belgium: OJ of the European Union. Retrieved June 2015, from http://eur-lex.europa.eu/legal-content/EN/TXT/PDF/?uri=CELEX:32012L0027&from=EN
9. Roberts, S. (2008). Altering existing buildings in the UK. Energy Policy, 36 (12), 4482–4486. https://doi.org/10.1016/j.enpol.2008.09.023.
10. Lemmet, S. (2013). *Buildings and climate change*. UNEP.
11. Ministry of Economy. (2014). *Pravilnik o sustavu za praćenje, mjerenje i verifikaciju ušteda energije. [Methodology on measurement and verification of EE savings] National Gazette, 1368(127/2014)*. Retrieved from http://narodne-novine.nn.hr/clanci/sluzbeni/2015_06_71_1368.html.
12. City of Zagreb. (2016, March 1). *ZagEE*. Retrieved from http://zagee.hr/?page_id=547&lang=en.
13. Ministry of Economy, (2014, July 30). *The third National Energy Efficiency Action Plan for the 2014–2016 period. 35. (E. Union, Trans.)*. Zagreb, Croatia: National Gazette. Retrieved 06 2015, from http://ec.europa.eu/energy/sites/ener/files/documents/2014_neeap_en_croatia.pdf.
14. Economidou, M., Atanasiu, B., Despret, C., Maio, J., Nolte, I., & Rapf, O. (2011). Europe's buildings under the microscope: A country-by-country review of the energy performance of buildings. Buildings Performance Institute Europe-BPIE.

Technological Quality in Process Innovation for Renewable Energy Buildings

Consiglia Mocerino

Abstract The building sector is one of the main indicators of environmental and economic sustainability because in the European Union (EU), it is responsible for around 40% of air pollution and 36% of final consumption.

Therefore, new building processes are aimed, above all, at environmental requalification with reconversion of the built environment, focusing on technological quality to achieve a 30% improvement in energy efficiency by 2030.

The objectives are process innovation and energy efficiency in construction, in accordance with the latest European Commission (EC) regulations and decisions, with achievement of a 35% share of energy consumed in the EU being obtained from renewable sources.

The performance capacity of the systems and products, with durability of the components, is aimed at building quality in accordance with the International Organization for Standardization (ISO) standard 8402 and ecosustainable control of energy production, with incentives for companies adopting smart energy solutions.

Innovative methodologies involve the application of efficient energy systems through the use of distributed generation, cogeneration and trigeneration with consumption control by sophisticated digital systems.

This chapter highlights new projects integrating design of efficient plants and new models for building with efficient systems and self-consumption of electrical energy for consumers and participation in the renewable energy community, with use of concentrated solar photovoltaic technology for energy self-production, solar thermal systems, and more.

These projects integrate passive systems and efficient double- and triple-skin architectural envelopes; sustainable and intelligent heating, ventilation and air conditioning (HVAC) systems; and smart materials with ecodesign. The challenges are environmental protection, reductions in CO_2 emissions, and containment of global warming to 2 °C, with innovation in new building processes that highlight technological quality in architecture, using renewable energy sources.

C. Mocerino (✉)
Faculty of Architecture, Sapienza University of Rome, MIUR, Rome, Italy

P. Bertoldi (ed.), *Improving Energy Efficiency in Commercial Buildings and Smart Communities*, Springer Proceedings in Energy,
https://doi.org/10.1007/978-3-030-31459-0_6

1 Introduction

Technological quality in energy-efficient buildings, using Renewable energy resources, involves many aspects of the environmental, social and political social sectors, aimed at strategies for a new building process. With a view to competitive economic growth and as an indicator of progress and technological development (in which construction is a predominant factor) for the purposes of Social sustainability, the aim is to achieve greater habitat performance. This mainly stimulates the search for high quality of both products and components of sustainable and efficient buildings, from an energy and environmental point of view, so renovation is entrusted to the complexity of the factors that synergistically define the technological quality of an architectural building, indicating, above all, the overall energy performance of the building with suitable levels of indoor air quality and ventilation, and relative lighting efficiency, both natural and artificial. Additional factors include Thermal–acoustic insulation levels; the solar approach; safety of the various components of the technological system; innovative materials; renewable energy technologies in Heating, ventilation and air conditioning (HVAC) systems; etc.

For these purposes, the new building process requires greater flexibility and cooperation of professional figures that can be interoperable with the support of the new Building information modeling (BIM) design and experts designing strategies with the adoption of new technologies in line with national and international programs and regulations that promote Energy efficiency in building, focusing above all on the commercial, office and tertiary sectors—hence, the practice of using clean technologies that increase the new energy efficiency markets, with all-green building products, innovative nanotechnologies and intelligent materials that exploit, above all, renewable resources.

Within the new building practice, the climate factor represents the focus for the adoption of new strategies for Sustainable development, which are aimed at the redevelopment of environmental energy and new achievements, through the application of innovative technologies that reduce CO_2 emissions. In fact, both anthropic action (with the use of fossil fuels and changes in land use) and buildings (which are responsible for about 40% of air pollution in the EU, and 36% of final energy consumption) are the main causes of greenhouse gas emissions, with the consequences of global warming; melting of glaciers; rising sea levels; the risk of flooding; and risks to water quality, human health, territory, agriculture, forestry, energy and tourism.

The objectives are economic growth with decarbonization and use of low-carbon technologies, according to the International Energy Agency (IEA), as highlighted in the World Energy Outlook (WEO) Special Report 2017. The polluting emissions of the construction sector are to be reduced by almost 60% by 2040 in comparison with present-day levels, mainly through a decrease in the use of coal and, to a lesser extent, oil, according to the IEA and the International Institute for Applied Systems Analysis (IIAS), as mentioned in the WEO Special Report 2016 (Fig. 1).

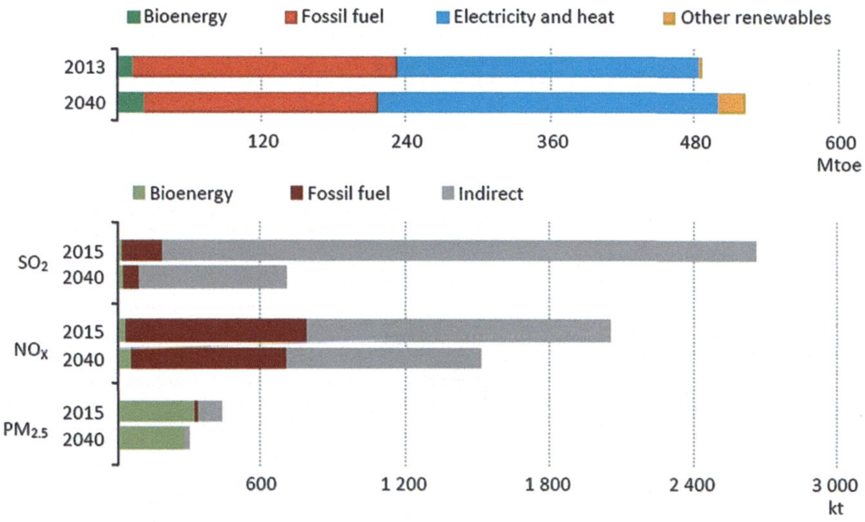

Fig. 1 Pollutant emissions from energy demand in buildings in the European Union construction sector [1]. *Mtoe* millions of tonnes of oil equivalent, NO_x nitrogen oxides, $PM_{2.5}$ particulate matter with a diameter of 2.5 μm or less

Another objective is the implementation of the European Ecodesign Directive, which reduces the average emissions of particulate matter with a diameter of 2.5 μm or less ($PM_{2.5}$) from solid fuel boilers. There will be a gradual increase in the use of Bioenergy in the construction sector (which will expand to almost 50% by 2040, especially for space heating); the building-related proportion of nitrogen oxide (NO_X) emissions related to energy will double to 17% by 2040, and the emissions will be comparable to those resulting from the production of electricity (which are currently about double). It is considered that the provision of safe, economical and modern energy to all citizens is essential for poverty reduction and economic growth. The first Sustainable development goal (SDG) is access to modern energy, which is a necessary condition to alleviate poverty, as in middle- and high-income countries. To this are added political and regulatory strategies to increase the circular economy, such as the Ecodesign Working Plan in the European Union (EU) [2], with the objectives of Energy efficiency and a new process for products and materials, considering their entire life cycle. Furthermore, in accordance with the recent agreements between the Council and the European Parliament, on the new European Directive 2010/31/EC-EPBD (Energy Performance of Buildings Directive) concerning the energy performance of buildings, energy savings targets aimed at renovation of existing buildings and new building by 2050 are highlighted. In the EU, additional plans, including building renewal, are established and indicators of results obtained (such as energy consumption limits per square meter, etc.) with medium-/long-term strategies for self-sufficient buildings such as NZEB (Nearly Zero Energy Buildings). Targets have been set, aimed at an increase of 35% in energy efficiency by 2030 (Fig. 2), with 35% of total energy consumption being

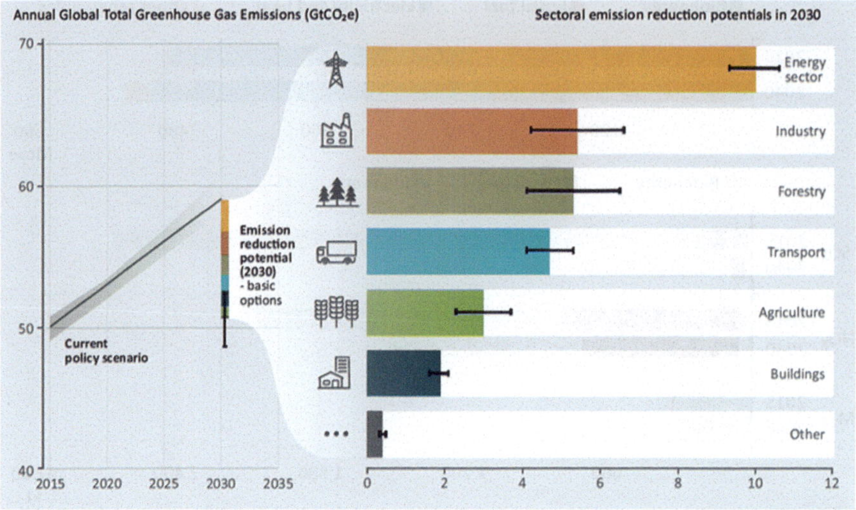

Fig. 2 Total emission reduction potentials, in comparison with the current policy scenario, in 2030. The potential global total reduction in greenhouse gas emissions ($GtCO_2e$) in 2030 is 33 $GtCO_2e$/year (uncertainty: 30–36 $GtCO_2e$) [3]

supplied from renewable sources (12% in transport), environmental protection with reductions in CO_2 emissions, containment of global warming to 2 °C and a reduction in greenhouse gas emissions of at least 40% in comparison with 1990. In addition, the Smart Readiness Indicator (SRI) highlights the performance requirements for buildings, related to intelligent systems and self-production, self-control and automation, with the ability to adapt to users' needs.

According to the United Nations Environment Program (UNEP) the potential reduction in global total greenhouse gas emissions ($GtCO_2e$) is about 18.5 $GtCO_2e$ by 2030 (range 15–22 $GtCO_2e$) in six energy categories: solar energy, wind energy, energy-efficient appliances, energy-efficient passenger cars, afforestation, and stopping deforestation.

In the EU, of all of the acclaimed energy strategies to promote renewable resources, we focus, above all, on energy self-consumption in buildings with installation of storage systems without expense burdens; on an increase, by 2025, in intelligent buildings with technological feasibility; and on economic development of buildings in which sensors for control of artificial lighting and indoor ambient temperature are applied. Moreover, an uniform power threshold of 70 kW is established for the inspection devices of Heating, ventilation and air conditioning (HVAC) systems. In addition, for redevelopment of the built environment and energy, we highlight innovative regulations, including Italian National Unification Agency (UNI) EN 16883:2017 [4], the new standard for the improvement of energy performance in historical buildings, and more. These solutions are part of a strategic plan in which climate change is challenged, but they push the environment and the Social economy as indicators of development, well-being, and competitiveness, with demanding new models for building processes within sustainable territorial reorganization and Technological innovation in accordance with the European Directive 2010/31/EU on the energy performance of buildings and Energy efficiency strategies, and in accordance with European Energy Efficiency Directive (EED) 2012/27/EU. Strategies for new energy models with solar concentration and storage systems, and mixed models for centralized production from cogeneration and trigeneration sources for the supply of thermal, electrical and refrigerant energy are among the most acceptable for district heating and district cooling. Photovoltaics can be used in buildings in different types of enclosures, including Building-integrated photovoltaics (BIPV), whose market has indicated a Compounded annual growth rate (CAGR) from 2014 to 2020 of 39% [5] and accounts for 1–3% of the total photovoltaic market worldwide, with about 200 different technologies. The IEA foresees that 16% of solar power production and 11% of solar thermal electricity will be produced by photovoltaics, with a total of 27% of electricity being produced from solar energy by 2050. In addition, there will be dynamic, adaptive, interactive, double- or triple-skin casings, and integration of passive systems into the energy systems of NZEBs in the building optics of the community of renewable energy and smart energy, which will also provide incentives for businesses.

Technological quality is specified by the International Organization for Standardization (ISO) standard 8402 [6] and UNI standard 10838 for new workspaces with intelligent automation systems and clean technologies, smart buildings for coworking, smart working, and energy efficiency in innovative models for construction types using off-site construction, prefabrication, and more. Finally, with UNI EN ISO 14040 it is established that also in the building industry the life cycle approach must be applied in analysis of production; in fact, Life cycle impact assessment (LCIA) defines life cycle phases through the description of borders that coincide with the production system, considering the life cycle of a product from its creation/extraction (e.g., from the quarry) to its disposal.

2 Energy Models in Housing Quality

In the new building process practice, performance requirements for components and materials are highlighted, giving the system/building the quality of a habitat in which all of the integrations of ecodesign and plant/energy are distinguished, focusing on renewable sources. The quality defined by the set of services or a product capable of satisfying the needs is in accordance with UNI 10838, based on ISO 8402, and distinguished in different functional, spatial, environmental, technological, technical quality, operational, utility, and maintenance aspects.

In addition to the identification of Energy efficiency requirements, the quality control system also achieves economic efficiency through the possibility of eliminating the causes of inefficiency of performance, together with verification of conformity of a product or of a process through operational activities and not management. This control is integrated into the design of the architectural structure, achieving satisfying results in the name of innovative construction and communication technologies, such as Information and communications technology (ICT) and BIM, which, in particular, improves the design performance in all phases of the building process, coordinated by a team of experts and designers. To this end, efficient organizational models for a new building process are identified in which the subjects interact differently and to which the different forms of procurement correspond. In this way, they realize the interoperability of the different phases of the process, through operational models, from planning and designing to implementation, management, maintenance, and up to decommissioning, in project activities that interface with customers.

Thus, we can achieve the best objectives of innovation, Energy efficiency and economy in the building sector, which accounts for about 40% of air pollution, and we focus mainly on office buildings and the service industry, in addition to the redevelopment of the built environment in a territorial reorganization program in which emergence issues of seismic and hydrogeological risk frequently stand out. Therefore, we can achieve quality construction, intended as the capacity to provide systems and products, and relative durability of components, with environmentally sustainable control of energy production and building automation with smart energy solutions and with incentives for companies. For this purpose, many of them have made the Circular economy and the cradle-to-cradle approach the bases for their business, hence offering sufficient data for early estimates by the Cradle 2 Cradle Network (C2CN) European platform.

The result is input from a variety of organizations, working together as customers and suppliers, in various forms of communities and networks, and in formal technology collaboration.

Furthermore, the overall quality of the construction highlights the application of energy technology with the use of renewable resources and ecosustainable materials in green building and their life cycle. According to UNI EN ISO 14040, the life cycle approach must be applied in the analysis of production in the building industry. In fact, it is the Life cycle impact assessment (LCIA) that defines phases of the product life cycle, through description of borders that coincide with the production system and consider it from its creation/extraction (e.g., from the quarry) to its disposal. The innovation in the standards of cost management in relation to energy consumption is significant and is linked to the policies of process and product strategies aimed at such quality solutions, with the development of Social sustainability of low cost and low profit, environmental sustainability and facility management, and public and private partnership.

The strategies of further models, in accordance with European Directive 2010/31/EU, EPBD recast [7], on the energy performance of buildings, are identified above all in the reduction of costs, in the achievement of NZEBs in 2020 with optimization of costs and in important requalifications. Hence, among the performance requirements for the Buildings Other than Dwellings specification, we highlight proposals with low transmittance values ($U = \text{W/m}^2\text{K}$) in roofs, floors and walls; elimination of thermal bridges; optimal values of air permeability; reduction of the window U

value, with reduction of transmittance of solar energy; advanced services and lighting specifications; and more.

Also, there are Building energy ratings (BERs)/advertising/display energy certificates and NZEB 2018 for buildings owned and occupied by public authorities, including the Office of Public Works (OPW), Local authorities (LAs), the Health and Safety Executive (HSE) and more. HVAC services should be in accordance with the Technical guidance document (TGD) heating efficiency (heating and hot water) proposal, with 91% gas boilers and with a high Seasonal energy efficiency ratio (SEER) value of 4.5, with savings in electricity consumption (electricity bills); automation of daylight and occupancy lighting controls, with 65 lm/circuit watt for lighting; and more. For important redevelopment, a percentage of over 25% of the building envelope renovation is established, indicating the modalities and providing measures for restructuring of inefficient heating, cooling and artificial lighting to optimize costs. With Directive 2012/27/EU on energy efficiency, a common system of measures is established, including "smart" meters for electricity and gas, to launch energy efficiency within the EU, with the aim of achieving 20% adherence to the minimum requirements by 2020. The building is intended as a set of housing, equipment, operation and maintenance, with the aim of potential development of Energy efficiency and reduction of CO_2 emissions and local pollutants for Sustainability. Audits include the energy audit with obligations for large companies and are performed by accredited bodies, qualified experts and energy management systems every 4 years.

Among the energy models using renewable sources are solar concentration, with the use of photovoltaic energy for self-production, and solar cooling systems with mixed models for centralized production from sources of cogeneration and trigeneration for the supply of thermal, electrical and refrigerant energy, which are among the most acceptable for district heating and district cooling. In particular, solar photovoltaics and wind energy (worldwide, 75 countries and 29 states are providing renewable energy with a premium price or specified price per kilowatt hour) are distinguished [8]. The costs of electricity from these energy technologies have already decreased in comparison with the electricity produced from fossil fuels [9]. In addition, storage systems without expenses are involved in models for buildings with efficient systems for electrical energy self-consumption, for heating and cooling control with intelligent technologies, for innovative and Light-emitting diode (LED) lighting systems for energy saving, and more. In these strategies the role of utilities, in support of local administration in energy efficiency and in the fight against climate change, is effective.

3 Envelope Performance with Efficient Technologies

The architectural configurations of different design models highlight both environmental and Energy efficiency, as a result of the new building process praxis in which integration of the architectural conformation and of both active and passive energy systems occurs. Thus, the goal of Technological innovation is to highlight the different types of envelopes with high performance and according to their intended use.

In office, commercial, administrative and tertiary buildings, in general, functional requirements for high standards of service provision, products and users in spaces for coworking are related to lighting with energy-saving installations in various workstation spaces and brainstorming spaces, the efficiency of connected building automation services, air quality, and more.

The Demanding–performance approach, with technoconstructive feasibility related to the systems and products, in their life cycle, leads to the development of new building process models that promote the application of sustainable and intelligent prefabricated off-site systems, in parametric modeling (Figs. 3 and 4). In fact, these innovations focus on Computer numerical control (CNC) technologies, toward the transformation of the process of the "Third Industrial Revolution" or "Industry 4.0" in which information technology (IT) integrates with the mechanical instrumentation of production.

These Computer numerical control (CNC) technologies are devices for automating milling operations with coded instructions sent to an internal computer, whereby all components and materials are cut and processed to precise specifications, facilitating timing and reducing construction costs, since they are produced at the factory and transported to the site, where they are assembled dry, without the use of mortars,

Fig. 3 Glass panels on the new Paris Courthouse, designed by Renzo Piano Building Workshop (RPBW Architects), 2017 (Photo: Consiglia Mocerino)

Fig. 4 Photocatalytic cement panel facade on the Italy Pavilion, Expo Milan, designed by Michele Molè, 2015 (Photo: Consiglia Mocerino)

Fig. 5 Off-site construction, using wood panels and prefabrication [10]

for the realization of the off-site prefabrication of the building (Fig. 5). Different materials are used, such as cement, glass, wood, metal, and more.

The praxis follows the design of efficient building systems that integrate with intelligent building automation systems using ICT technologies to operate all of the services and application of energy systems with renewable sources.

Among these systems, the energy-efficient lighting project manages to stimulate the adoption of efficient equipment with minimal labeling and energy performance. In fact, the Human visual performance depends on the quality and quantity of light; to maximize the use of natural light, the installation of work lights takes place where it is needed. For this reason, various systems of daylighting are adopted to take into account the compass directions, among which the most advantageous are south-facing windows for direct light and high summer temperatures, with north-facing windows for regular daylighting and natural light, avoiding the high-beam summer light; those in the east or west, while providing good natural daylight, can cause glare.

In the interiors of office buildings and the tertiary sector, in general, energy-efficient appliances are distinguished by indoor lighting technologies using incandescent lighting (without a ballast, with intense light that is dimmable, and with instant control) and fluorescent lamps (which require a ballast, with efficacy of 30 to 110 lm/W), lasting about ten times longer (7,000–24,000 h). These lighting systems include intelligent appliances (with on–off toggle switches, occupancy sensors, photosensors, timers) with indoor strategies (to maximize the use of daylighting, with installation of fluorescent light) and outdoor strategies (using fluorescent, High-intensity discharge (HID) lamps, and more).

The consumption of electricity has been much reduced with use of new lighting technologies and lighting efficiency standards in commercial buildings in the USA.

According to a report from the Energy Information Administration's (EIA's) Commercial Buildings Energy Consumption Survey (CBECS) [11], these include standard fluorescent lighting (which is more energy efficient and has a longer lifespan than incandescent bulbs), Compact fluorescent lamps (CFLs), incandescent lighting, and HID lighting, used in open layouts, storage areas with high ceilings, and box retail stores. For very bright light, the halogen category of incandescent bulbs provide high-quality light with increased Energy efficiency, and LED lamps offer high efficiency with low energy consumption for bollard lighting, etc. Other innovative lighting types are used with BIM, 3DMax and VRay systems.

For the purposes of optimal lighting from both artificial and natural sources in buildings (especially for offices), and for good Thermal insulation, with good air quality and constructive and environmental safety, it is necessary to realize energetically efficient envelopes with the adoption of efficient construction technologies, through a BIM design that considers many qualitative and sustainable parameters.

In fact, in process innovation, designers must consider several main aspects, including the climate, the shape of the building and its orientation with respect to the context, and the energy requirements, for which it is necessary to consider the classes of needs in relation to quality requirements for design and quantity, with related control indicators for compliance with the requirements and the satisfaction of the occupants.

The classes of needs include the quality of internal and external and related to connected to them, related to different requirements, including materials, air quality, noise reduction, thermohygrometer well-being, and the quality of services for related control requirements of building/plant systems. Moreover, there is a focus on Environmental/climatic data and related resource needs concerning classes of atmospheric and electromagnetic pollution requirements of outdoor spaces; and heating, cooling and natural ventilation requirements, including passive system requirements, night mass ventilation, and natural ventilation generated by technologies with construction of chimney effects, and more.

Consideration of reductions in Environmental loads is also necessary in relation to the requirements regarding atmospheric pollution, waste management, and for the purpose of control, above all, of emissions, associated with other contextual and environmental integration needs in relation to external spaces, protection of the landscape and of the territory, and requirements regarding electromagnetic pollution, noise, etc.

Lastly, there is a distinction between the need to save resources in terms of the use of building materials, renewable energy, the control of solar thermal contributions, mechanical ventilation, the production of electricity and the quality requirements of services corresponding to control needs for buildings and installations that highlight the technological quality of the building organism.

Therefore, the descriptive[1]/performance indicators are mainly related to anthropic activities with respect to the explication of a given phenomenon, which, according to european environment agency (EEA), quantify the state of the environmental factors,of the health safety, and more.

The different performances of Ecosustainable models are integrated with those of new energy models aimed at the use of electricity through the exploitation of renewable resources and recourse to storage systems, to guarantee the thermal and hygrometric comfort of the confined spaces and the containment of energy consumption, through energy models with energy characteristics and innovative technologies. In fact, the typologies are distinguished, above all, by environmental control of conservative, selective, regenerative, and advanced bioclimatic types, adaptive intelligent dynamics, etc.

For environmental control, conservative enclosures are aimed at the reduction of heat losses with large masonry and few lights, while selective typologies involve the use of large, transparent walls for lighting and passive heating (double skin, triple skin, and more).

Instead, for regenerative envelopes, we highlight plant systems with an envelope as a barrier between inside and outside. For advanced Bioclimatic typologies, environmental control occurs with the interaction between the context (exterior space changes, etc.) and the building for energy flow management.

In these typologies, together with design and intelligent technologies, the use of innovative, adaptive materials that give efficiency and dynamism to building envelopes is fundamental. In fact, dynamic enclosure components and systems are identified with Phase change materials (PCMs), smart windows, chromocene materials and integration with HVAC plant Thermal mass activation (TMA), Earth coupling (EC), Dynamic insulation systems (DIS), and Advanced integrated facades (AIFs).

Double- and triple-skin facades are made with special deformed thermoinsulating glass panels[2] with Thermoisolating properties (with air encapsulated between

[1]The descriptive indicators come from the indicators in the World Bank core set, the United Nations Conference on Sustainable Development (UN-CSD) of the USA, and the structural indicators issued by the European Council and the Organization for Economic Co-operation and Development (OECD), among the main international institutions.

[2]In multilayer insulating glass, at least two sheets of glass, sealed only along the edge, enclose a

Fig. 6 Double skin with photovoltaics on the new Paris Courthouse, designed by Renzo Piano Building Workshop (RPBW Architects), 2017 (Photo: Consiglia Mocerino)

two glass sheets) and thermo-higrometric well-being inside dynamic enclosures, with environmental control through regulation of the flows of light, air and water from the outside to the inside and vice versa. They react and adapt effectively and efficiently to climatic variations for comfort and energy/environmental well-being.

Energy-efficient facades of buildings are made with a skin covered by Photovoltaic modules (Fig. 6) using the latest generation of polycrystalline silicon, monocrystalline silicon and thin film, with a low environmental impact, which occupy a smaller surface with the same power generation, exceeding 3 kW, from small plants, and are amortized in a few years at a low cost.

There is a trend toward use of multicrystalline silicon, as per the research of the Fraunhofer Institute for Solar Energy Systems (ISE), which converts 21.9% of incident solar radiation into electrical energy. Therefore, we highlight the need for energy savings in models for mainly transparent casings with passive facades and breathing windows, with intelligent ventilation that blocks solar radiation (which is reflected by the blinds) and is aimed at improving thermodynamic and acoustic performance.

Generally, ventilated facades are based on two layers that form the first skin or selective glass covering or another material installed in the supporting structure of different types, including metal, wood, cement, ceramic, bricks, etc., and it is

hermetically sealed space between the panels (SZR Scheibenzwischenraum o LZR, 8/16 mm). They usually use float glass, single-layer safety glass (ESG) or compound laminated glass (VG), reinforced glass (TVG), or laminated safety glass (VSG). The insulating glass panel is independent of the frame, constituting an autonomous system.

anchored to the load-bearing structure of the building. This first coating, which closes the structure of the building, contains openings equipped with intelligent systems of solar shading with installation of sensor systems, and provides the best options to manage the interaction between environmental fluctuations that affect the building, exterior and interior spaces, and indoor comfort.

Therefore good thermal performance capacity and relative cost reduction of HVAC systems (heating, ventilation, air conditioning) are obtained.

The first skin can efficiently control solar energy inside the building, and it interposes a space, a buffer zone, or a natural ventilation cavity allowing inspection and maintenance of the main facade of the structure. Thus, types of double-skin glazed facades, etc., can achieve the objectives of ecocompatibility, energy efficiency and Technological quality (Fig. 7), adhere to the same production policies as BIPV buildings, with curtain walls, roofs, windows, walls, etc., with energy integrated by Photovoltaic modules, mainly flexible thin film, highlight the quality of the technologies and materials of an integrated ecodesign and energy plant. In particular, the construction of these types of Building-integrated photovoltaics (BIPV) buildings is rapidly growing in North America and the Far East, and they are widely distributed in European markets with the promotion of feed-in tariffs, launched by leader in photovoltaics, including Saint-Gobain, Lumeta, Applied Solar, AGC Solar (with SunEwat XL safety glass modules for Europe and Sunjoule modules for the markets in Japan), Schott, Dyesol, Dow, and more. They have been the subject of numerous research studies for the application of Renewable energy sources (RES). In fact, for better production and distribution of BIPVs, price competitiveness with high performance is fundamental, which depends, in large part, on optimal orientation of the building system, on the life cycle and duration of the materials, on the quality of the product, and on resistance to atmospheric agents. So, what we mean for energy performance is the level of energy consumed per year or that can meet the energy

Fig. 7 Double glass skin on the Teca, part of the new Rome/EUR Convention Center, designed by Studio Fuksas, 2016 (Photo: Consiglia Mocerino)

needs (summer and winter air conditioning, lighting, hot water production and ventilation). Specific standards and building codes for these particular photovoltaic buildings have been developed, specifying building materials, architecture, electrical safety, production of standard modules, and more. Worldwide, these standards include those issued by the International Electrotechnical Commission (IEC), ASTM International, the EU (the Eurocodes), UL LLC, and more.

Generally, the operation of photovoltaics, for the purpose of electricity production in buildings, is that of a type of photovoltaic that is connected to a grid and provides a global energy benefit, without batteries, thus reducing air pollution rates.

Stand-alone, off-grid, self-employed Photovoltaic technology is applied in envelopes where there is a requirement for supply to remote areas or a substantial supply need not served by the national grid.

These technologies produce electricity from solar energy, which is produced by Electromagnetic energy resulting from the merging of hydrogen present in the sun.

Photovoltaic on-grid systems produce, convert and transfer solar electricity to the 230/400 V network for residences and factories. Photovoltaic off-grid systems, also known as Solar autonomic or island systems, work autonomously in a 12/24/48 V environment, supported by external sources from wind/water/electric power grids.

The efficiency of photovoltaic cells and modules results from the photovoltaic technology and is defined by the ratio between the amount of incident solar energy and the amount of energy produced (or, in the case of solar collectors, thermal power).

A solar panel with 14% efficiency will produce 140 W of electricity from 1 m^2 (or 140 Wh/h from 1 m^2). Usually, photovoltaics are also integrated into photovoltaic brise-soleil systems in building facades (Figs. 8 and 9). BIPVs are controlled by building automation sensors.The ratio between the amount of incident solar energy and the electricity produced, defines the efficiency of both cells and photovoltaic modules, which depend mainly on photovoltaic technology.

Fig. 8 Building-integrated photovoltaics on the Service Public de Wallonie, Namur, designed by Atelier d'Architecture Thierry Lanotte, using ISSOL glass, 2017 [12]

Fig. 9 Building-integrated photovoltaics on the new Paris Courthouse, designed by Renzo Piano Building Workshop (RPBW Architects), using ISSOL glass, 2017 (Photo: Consiglia Mocerino)

4 Examples of Best Practice

Environmental energy strategies are highlighted in many new and upgraded interventions [13], related to different climatic/environmental and territorial areas, where high performance of Energy efficiency, technology and architectural quality are subordinated to an integrated building design/system, using parametric processing with application of intelligent systems, for different uses. They are distinguished by dynamic envelopes with quality certifications. This chapter highlights three interventions of buildings for offices and a museum redevelopment: the headquarters of Métropole Rouen Normandie in Rouen, France, designed by Jacques Ferrier Architecture (JFA); the redevelopment of the Borghese Gallery in Rome, Italy, designed by Prof. Ing. Livio de Santoli; and the new ACCA Software headquarters, an Industry 4.0 smart factory in Bagnoli Irpino, Italy, designed by Prof. Arch. Francesco Bruno.

4.1 Headquarters of the Métropole Rouen Normandie, Designed by Jacques Ferrier Architecture (JFA), 2017

In a regeneration intervention in the Flaubert Ecodistrict of Rouen, a new building has been constructed for Métropole Rouen Normandie, with a colored glass facade inspired by the impressionist paintings of Claude Monet. This contextualized building, on the bank of the river Seine, re-evokes, with its outline, the pre-existing industrial buildings that were rebuilt on the right bank. The dynamic architecture includes offices, open space, and coworking and sharing spaces, with about 560 users. There are two volumes with a central light space and a central patio, ground floor reception, and offices distributed in the upper parts. The load-bearing structures are in cement and steel, with glass casing on the metal structures, and they are energy efficient and sustainable. The Building-integrated photovoltaics (BIPV) has a double skin of about 5,385 multicolored glass panels, with photovoltaic panels on the facade (on the south side) (Fig. 10a) and on the roof (Fig. 10b), from the firm ISSOL. The cost of the glass was about one twentieth of the total cost of investment in artistic structures.

The predominant design indicators in these constructive Technological innovations are:

- An interior central light atrium (Fig. 11); a double skin with a passive casing (passive thermal protection); photovoltaic panels on the roof, with high performance and using different colors.

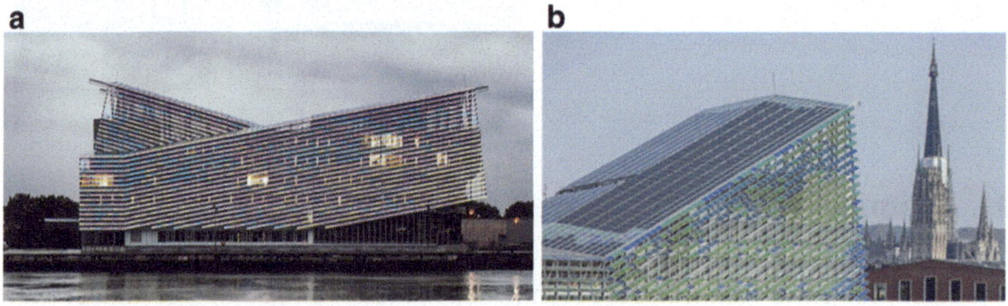

Fig. 10 Building-integrated photovoltaics on the headquarters of the Métropole Rouen Normandie, designed by Jacques Ferrier Architecture (JFA) [14]. (**a**) Facade with colored glass photovoltaics. (**b**) Roof with colored glass photovoltaics

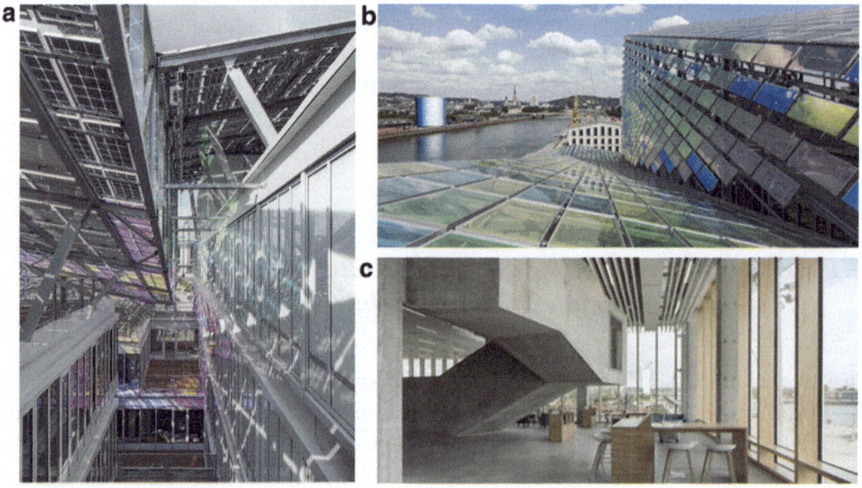

Fig. 11 Headquarters of the Métropole Rouen Normandie, designed by Jacques Ferrier Architecture (JFA) [14]. Double-skin atrium facade (**a**) and roof (**b**) with colored photovoltaic glass. (**c**) Open space with wooden interior design panels [14]

- Glass technology: Lamination of photovoltaic glass, with the insertion of a high-tech film (dichroic filter). This decomposes the light spectrum into different colors that are perceived as a function of the point of view and also generate color reflections. There is open space for smart work and finishes with a wooden design.

4.2 Redevelopment of the Borghese Gallery, Rome, Designed by Prof. Ing. Livio de Santoli

This energy regeneration project has been carried out the Borghese Gallery in Rome (Fig. 12), a building from the 1600s, located in the Villa Borghese. It has achieved Leadership in Energy and Environmental Design for Existing Buildings: Operations & Maintenance (LEED-EBOM certification and won a Best Practice Award—Public Heritage in 2017.

Sustainable interventions are aimed at energy redevelopment for Energy efficiency with improved energy performance and environmental requalification. In particular, UNI EN 16883:2017 contains the guidelines for the improvement of energy efficiency of buildings of considerable historical/architectural/cultural interest belonging to each period. They evaluate the reduction of energy consumption through improvement and performance choices and through investigation, analysis and documentation about the building.

Therefore, the interventions, through a preliminary evaluation phase (focusing on design credits) that guided the choices of the LEED-EBOM quality certification protocol, led to the improvement of the winter energy performance index, with a value of 320.8 kWh/m^2/year, reaching class A1 and primary energy requirements, for winter air conditioning, with 19% from renewable sources. The energy requali-

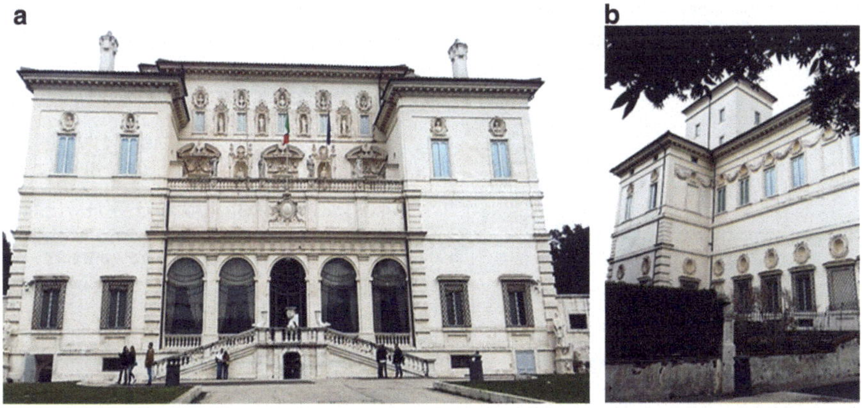

Fig. 12 Borghese Gallery, Rome. (**a**) Front facade. (**b**) Side facade (Photos: Consiglia Mocerino)

fication highlights the minimum performance credits both for ventilation and for indoor air quality, distinguishing itself in the following fundamental interventions:

- *Intervention A* (Fig. 13): Building automation with supervision system class B (UNI-2012) control of the air conditioning system. Credits for solution A: plant building system; measurement of energy consumption of buildings and advanced plant systems; Energy efficiency solutions.
- *Intervention B* (Fig. 14): Ground floor, in the ceiling/windows: installation of existing fan coils for external air intake.
- *Intervention C* (Fig. 15): Mariano Rossi Salon: plant 20×13 m^2, height 17 m. Solutions: Two dedicated air treatment units, in the air space and under the gallery's first-floor terrace. Air introduced into the rooms through nozzles to launch, expressed with grids in the upper areas of perimeter frames.
- *Intervention D* (Fig. 16): Entry and extraction of air through existing monumental fireplaces.

Credits for solutions *B*, *C*, and *D*: Minimum performances for internal quality, increase in ventilation.

- Installation of two polyvalent heat pump units (located in the Garden of Venus near the gallery) for the production of thermoconvector fluids.
- Credits for renewable energy sources.

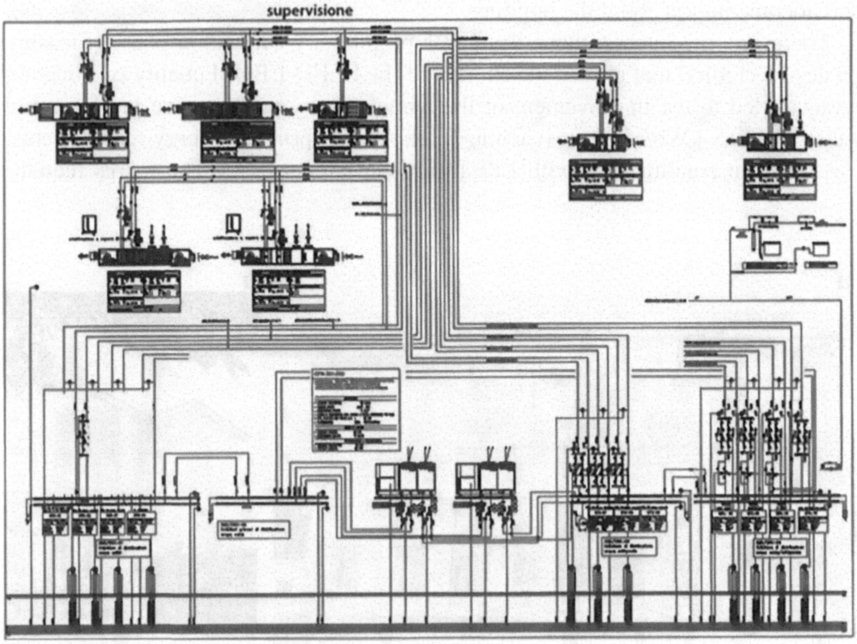

Fig. 13 Borghese Gallery, Rome. Intervention A: functional scheme of the air conditioning system, with control from the supervision system [15]

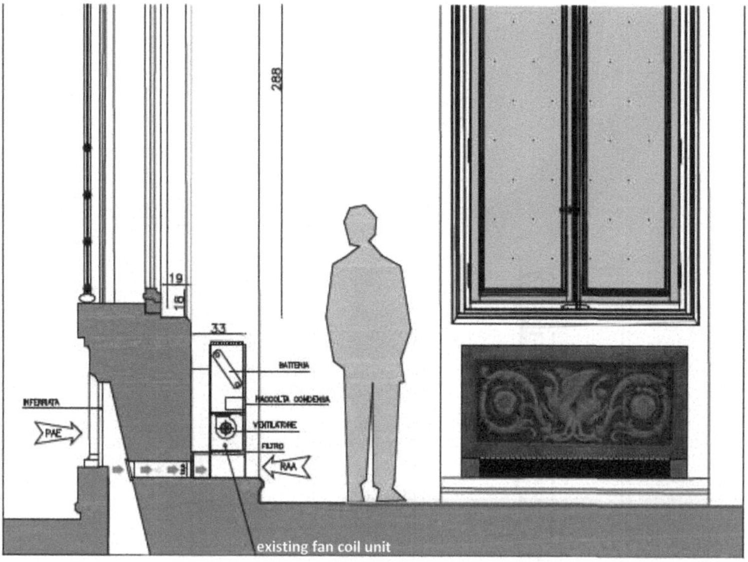

Fig. 14 Borghese Gallery, Rome. Intervention B: fan coil with external air intake [15]

4.3 New ACCA Software Headquarters, Bagnoli Irpino, Designed by Prof. Arch. Francesco Bruno

This building, built in 2016 in Bagnoli Irpino (AV) in the Area PIP (intended for production facilities), designed by Prof. Arch. Francesco Bruno, is state-of-the-art, zero-impact, sustainable, and self-sufficient architecture with energy resources from renewable sources including solar, thermal, photovoltaic and wind energy. The architectural work, with a reinforced concrete and antiseismic-bearing structure, highlights glazed and single-layer masonry walls using high-energy-performance bricks. It is contextualized (Fig. 17a) with the environment by means of a spatial articulation distributed in four parallel blocks and connected by a central block, with a glass envelope (Fig. 17b), and has been designed to house the headquarters of ACCA Software, a computer technology company that specializes in technical processing in the construction industry. The parametric design, using BIM design technologies, features efficient envelopes with a double-glazed skin; shielding systems for reducing solar radiation; functional solutions for indoor wellness, safety, comfort and savings; energy-open space; smart working with a "mind gymnasium" central flexible space; and a strategic work room.

The new ACCA Software headquarters was created to meet the needs of work and new entrepreneurship in the trend of smart factories using artificial-intelligence building automation and adoption of a Cyber/physical system (CPS) [17], which allows robots to interact efficiently and independently with the surrounding environment and communicate with it, related to spaces for smart working, open sharing commercial spaces, interoperable spaces, etc.

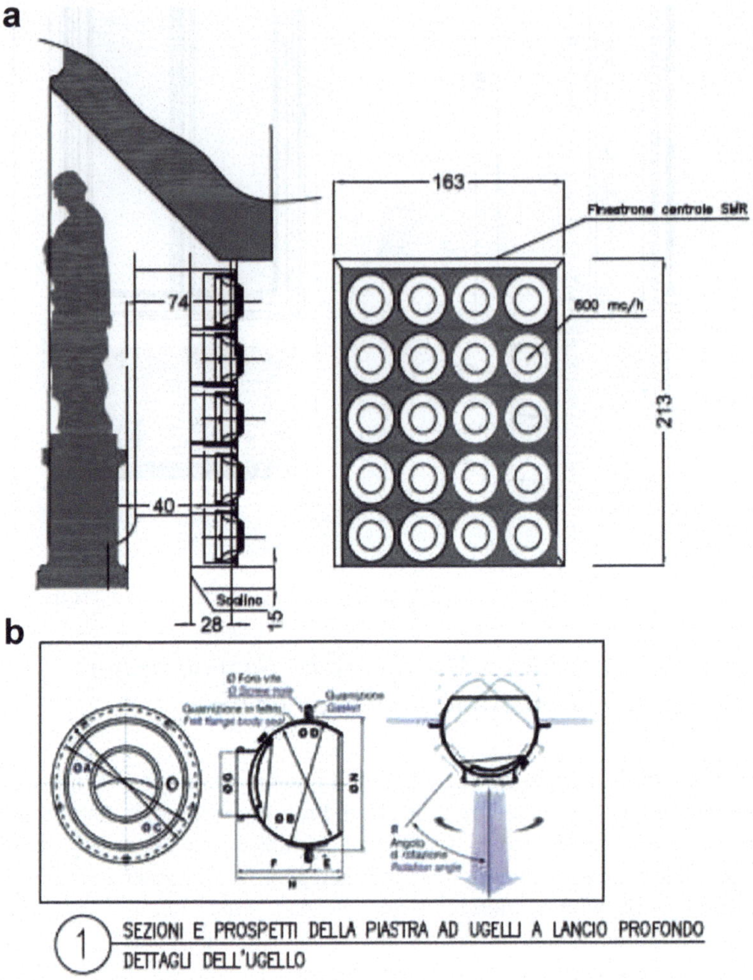

Fig. 15 Borghese Gallery, Rome. Intervention C: Mariano Rossi Salon [15]. (**a**) Deep-jet nozzles for air intake. (**b**) Plate with deep-jet nozzles—detail of sections and elevation

Internet of things (IoT) objects and Big Data are linked to the CPS of the ACCA Software headquarters in accordance with Industry 4.0, with the advantage of controlling all business processes in real time and performing simulations and predictive analyses through Big Data.

The following features stand out:

- *performance requirements.*
- Double-skin building envelope (Fig. 18a)
  - Energy class A4.

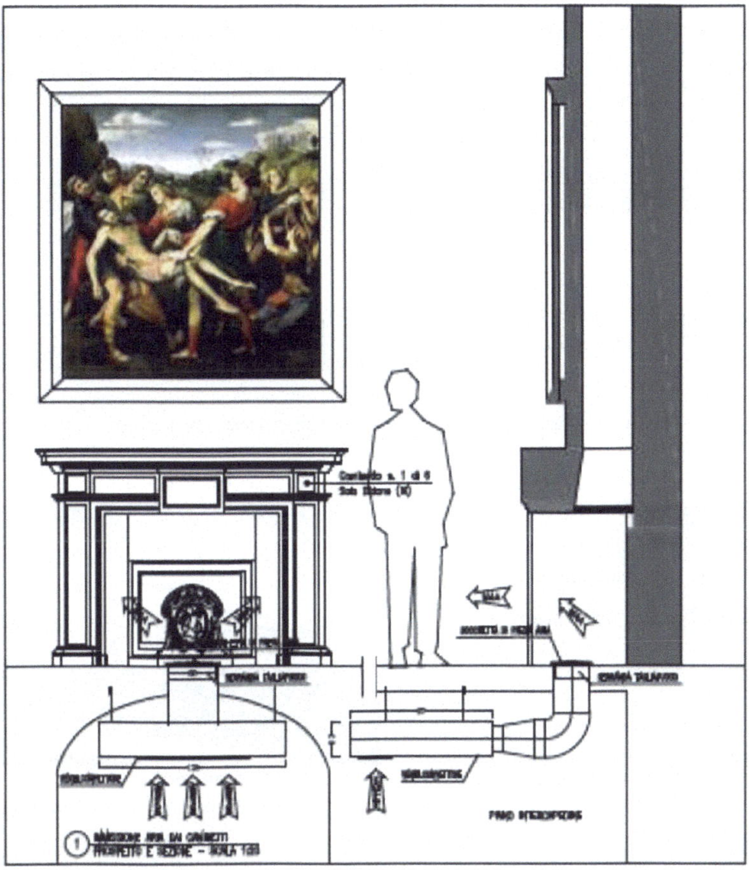

Fig. 16 Borghese Gallery, Rome. Intervention D: air intake in rooms through the use of six pre-existing monumental fireplaces [15]

- Global energy performance index, nonrenewable energy (EP_{gl}, nren) 32 kW/m^2.
- Global energy performance index, renewable energy (EP_{gl}, ren) 67 kW/m^2.
- CO_2 7 kg/m^2.

- *energy performance.*
- Energy self-sufficiency, using renewable energy sources: 200 kW, three wind turbines
  - Photovoltaic panels in the cornice structure from east to west (1100 m^2) and parking shelters (Fig. 18b).
  - Solar thermal system (swimming pool).
  - Shed cover, natural north-light axis to south photovoltaic integration.
  - Wind tower, recovery system, collection and reuse of rainwater.

Fig. 17 New ACCA Software headquarters, Bagnoli Irpino, designed by Prof. Arch. Francesco Bruno. (**a**) Perspective. (**b**) East–west section [16]

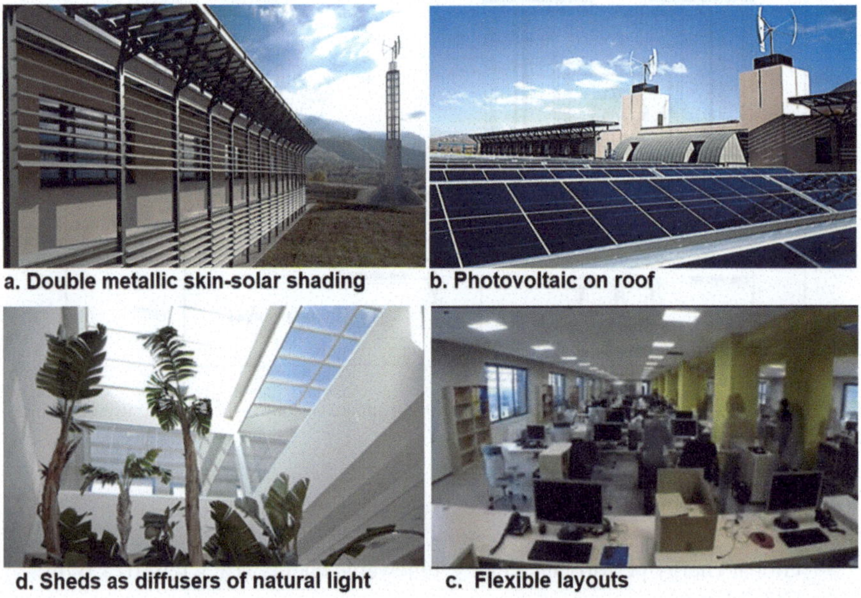

Fig. 18 New ACCA Software headquarters, Bagnoli Irpino, designed by Prof. Arch. Francesco Bruno. (**a**) Double-skin metallic solar shading [18]. (**b**) Photovoltaics on the roof [19]. (**c**) Flexible layouts [20]. (**d**) Sheds as diffusers of natural light [19]

- artificial lighting LED technology:
  - General lighting (ABSENT) recessed into an exposed ceiling—luminous efficiency 162 lm/W, analog dimmable.
  - Accent light (HALL LED)—luminous efficiency 100 lm/W for inputs, wide spaces, scenographic effects, etc.

  - Commission Electrotechnique Internationale (CEI) EN 60598-1, UNI EN 12464-1 standards.

- *building performance.*
  - Zero impact, covered area 8,000 sq.m on 30,000 sq.m.
  - Environmentally sustainable corporate headquarters, with flexible layouts for integration, sharing and coworking (Fig. 18c).
  - Natural lighting.
  - Use of natural materials and products throughout the entire life cycle of the building.

- *environmental wellness and comfort.*
- Natural lighting from diffusers to sheds in northern lighting coverage (Fig. 18d)
  - Indoor air quality, with controlled mechanical ventilation (CMV) through intelligent and automatic systems.
  - Heat recovery and thermal dispersion control.
  - Indoor zones with greenery.

- *technological and environmental quality.*
- Appearance, usability, comfort, safety, security.
- Fiber-reinforced base coat with a water-repellent finish.
- Main construction technologies: facilities in reinforced concrete with double-envelope high-energy performance
  - Rectified brick (Porotherm BIO PLAN 45 T – 0,11 with a Kl 7 system—fiber-reinforced base coat with a water-repellent finish made of dry mortar: Portland cement, polymeric fibers, additives, water-repellent material, classified sands) with mortar joints of 1 mm, elimination of thermal bridges, and increased energy performance; metal structure (distance from main structure: 80 cm) on west and east facades (solar horizontal and vertical shading).
  - False ceiling: acoustic panels (Knauf-EN ISO 11654-AMF Thermatex Alpha One Acoustic Range)—fireproof, thermohygrometric, antireflective, etc.

- Among the efficient building technologies, on the west and east facades of the working areas there is an innovative system with double skin in steel anchored to the main structure of the building, with a distance of about 80 cm and equipped with automated vertical and horizontal mobile screens, using the Orsogril system.

Detail (Fig. 19a–c): Metallic support structure with solar shading (east/west) with adjustable slats, automated according to the movement of the sun and the characteristics of the light (Fig. 19a):

- Horizontal (Orsogril-type panels, accessible also for maintenance of the structure and fixtures).
- Vertical (sliding and adjustable).

a. Prospectus b. Perspective c. Vertical Section

Fig. 19 New ACCA Software headquarters, Bagnoli Irpino, designed by Prof. Arch. Francesco Bruno. (**a**) Front elevation [21]. (**b**) Perspective [21]. (**c**) Vertical section [22]

The building is equipped with artificial-intelligence systems [23] and building automation for control and management of the thermoregulation systems, lighting, operation of current loads, outgoing and incoming communications of the building, comfort, security and anti-intrusion, and global building safety (fire protection, air quality, etc.).

These innovations realize energy savings, comfort and occupant safety in the following ways:

- Three-level multiprocessor system interconnected by serial buses: The actions to be performed are controlled by a central computer, after receiving all relevant information.
- The information is sorted by 12 intermediate processors (floor boxes)—one for each building block.
- Utilities interface with 111 field processors (three box types).

functioning:

- Supervision and control system for automatic thermoregulation (climate change, user needs, energy saving).
- Lighting system sensors and actuators (opening or closing control of sunshades as a function of solar radiation), electronic switches, etc.
- Control of current load operation and all communications systems, emergencies, audio–video sources.
- Security (anti-intrusion).
- Safety (safe operation of services, gas, fire, air, etc.).
- Real-time acquisition of all information (radiofrequency (RFID) identification, etc.), control/regulation of air conditioning systems.

5 Conclusions

Sustainability in buildings involves many related sectors, including professional skills, economic policies and social policies, focusing on Renewable energy resources and Energy communities. These indicate greater participation and awareness of the strategic solutions in masterplans and new programming for building interventions, redevelopment of the built environment, in accordance with UNI EN 16883:2017, and new construction by launching NZEBs, with a view to energy efficiency and economic sustainability/environmental quality in accordance with the regulations. The goals are a 35% increase in Energy efficiency by 2030, 35% of total energy consumption coming from renewable sources, and a reduction of at least 40% in greenhouse gas emissions, focusing on the SRI of building performance requirements, with intelligent systems and self-production, self-control, automation and adaptation to user needs. Use of Energy-saving construction technologies is needed, with a forecasted expenditure in both Western and Eastern Europe of $111.9 billion in 2026 [24], focusing on the building envelope and control of the building, lighting systems, HVAC and other systems, with related commissioning and installation. Furthermore, for energy efficiency, it is essential to adopt artificial-intelligence building automation and a CPS that allows robot machines to interact efficiently and independently with the surrounding environment and to communicate in smart working spaces.

From this perspective, quality in construction, in accordance with ISO 8402 and UNI 10838 for building quality, is emerging and is connected to a series of factors that flow mainly into the new building process, whose organizational models involve new figures in an interoperable practice [25], adopting smart technologies and systems that increase the performance of building interventions, with reference to incentive rules including European Directive 2010/31/EU on the energy performance of buildings and (EED) European Directive 2012/27/EU on energy efficiency in a building process of integration between architectural design and energy plants.

Thus, we are launching various intelligent technologies that exploit renewable resources, with the adoption of thin-film photovoltaic systems, the latest-generation ICT technologies, etc., integrated into innovative digital and CNC technology solutions, passive systems of double and triple skin building enevelopes, involving BIM design and innovative off-site construction and prefabrication types.

The practice mainly involves office, commercial and service buildings with applications of new energy technologies, aimed at durability of the products and components, and related quality control with certification and labeling systems in accordance with standards such as the European Ecodesign Directive and national and international standards. They are aimed at Sustainability and Energy efficiency targets with smart energy solutions, including incentives for companies, with inclusive processes in all countries for new planning and greater responsibility for human, social and environmental rights, with CO_2 reduction, productivity, and habitat quality. In particular, the EU takes effective protection for reducing emission by 80–95% by 2050 and for the provision of secure, affordable and modern energy for citizens' economic growth. Hence, new and intelligent methodologies have been devised that are efficient and sustainable in terms of the quality and impact of products, in accordance with UNI EN ISO 14040, which specifies that in the building industry the life cycle approach must be applied in the analysis of production, with Life cycle impact assessment (LCIA) evaluation defining phases of the life cycle through the description of borders that coincide with the production system, considering the life cycle of a product from its creation/extraction (e.g., from the quarry) to its disposal. The challenge is to achieve Technological quality in construction that aims to reduce CO_2 emissions with new systems for energy-efficient buildings, investing in innovations for the real estate market and adopting intelligent IT technologies and systems.

References

1. Energy and air pollution – World energy outlook 2016 special – IEA. OECD/IEA 2016, World energy outlook special report: Energy and air pollution, IEA. Retrieved from www.iea.org/t&c, https://www.iea.org/.../WorldEnergyOutlookSpecialReport2016EnergyandAirPollution. Accessed January 2018.
2. COM (2016) 773 final, Ecodesign Working Plan 2016–2019 – European Commission – Europa. Retrieved from https://ec.europa.eu/energy/sites/ener/.../com_2016_773.en_pdf. Accessed January 2018.
3. The emissions gap report 2017 – UNEP document repository home. Retrieved from https://wedocs.unep.org/bitstream/handle/20.500.11822/22070/EGR_2017.pdf. Accessed January 2018.
4. UNI EN 16883:2017, Conservation of cultural heritage – Guidelines for improving the energy performance of historic buildings. Retrieved from www.uni.com. Accessed January 2018.
5. Development of innovative educational material for building-integrated photovoltaics. Retrieved from http://www.dem4bipv.eu/bipv/bipv-market/. Accessed February 2018.
6. UNI EN ISO 8402:1995 Gestione per la qualità ed assicurazione della qualità. Retrieved from http://store.uni.com/catalogo/index.php/uni-en-iso-8402-1995.html. Accessed December 2017.
7. Directive 2010/31/EU of the European Parliament and of the Council. Retrieved from http://eur-lex.europa.eu/LexUriServ/LexUriServ.do?uri=OJ:L:2010:153:0013:0035:EN:PDF. Accessed November 2017.
8. Renewables 2017 global status report – REN21. Retrieved from www.ren21.net/wp-content/.../2017/06/17-8399_GSR_2017_Full_Report_0621_Opt.pdf. Accessed January 2018.
9. Renewable power generation costs in 2017 – IRENA. Retrieved from https://www.irena.org/-/media/Files/IRENA/.../IRENA_2017_Power_Costs_2018.pdf. Accessed January 2018.
10. Durapanel. Retrieved from https://www.durapanel.co.nz/about-us/. Accessed January 2018.
11. EIA – CBECS lighting report (May 17, 2017). Trends in lighting in commercial buildings. Retrieved from https://www.eia.gov/consumption/commercial/. Accessed January 2018.
12. Service Public de Wallonie. (2017). *ISSOL architecture BIPV, active facades & building skins. Technological constructions photovoltaic Brise Soleil.* Namur: Service Public de Wallonie. Atelier d'Architecture Thierrry Lanotte, http://www.issol.eu/nl/home-nl/. Accessed February 2018.
13. Mocerino, C. (2017). *TOP OFFICE Tecnologie intelligenti di riqualificazione.* Rome: Gangemi Editore S.p.A. International. ISBN:978-884923540-L.
14. JFA Jacques Ferrier. Project. Métropole Rouen Normandie 2017. Retrieved from http://www.jacques-ferrier.com/projets/metropole-rouen-normandie/#4;#7;#9;#15;#20. Accessed February 2018.

15. de Santoli, L., Perini, G. P., & Rossetti, S. Certificare la sostenibilità dei Beni Culturali-Il caso della Galleria Borghese. *AjCARR Journal, 43*, 8-marzo-2017. ISSN:2038–2723. Accessed March 2018.
16. Nuova sede ACCA, Bagnoli Irpino: Presentazione. Company Profile. ACCA software. Retrieved from https://ihc2015.info/skin/acca-acca.akp. La nuova sede ACCA-Concept, Prof. Renato De Fusco. Retrieved from https://www.acca.it/nuova-sede. Accessed February 2018.
17. Alur, R. (2015). *Principles of cyber-physical systems*. Cambridge: MIT Press. ISBN:9780262029117.
18. Nuova sede ACCA, un edificio in armonia con il territorio. Retrieved from https://www.info-build.it/progetti/nuova-sede-acca-un-edificio-in-armonia-con-il-territorio/. Accessed February 2018.
19. Nuova sede ACCA. Retrieved from https://www.acca.it/nuova-sede. Accessed February 2018.
20. Time-lapse del trasferimento di ACCA software nella nuova sede di Bagnoli Irpino, cover image 480 × 360. Retrieved from https://www.youtube.com/watch. Accessed February 2018.
21. Nuova sede ACCA, un edificio in armonia con il territorio. Retrieved from http://www.info-build.it/progetti/nuova-sede-acca-un-edificio-in-armonia-con-il-territorio/. Accessed February 2018.
22. La nuova sede ACCA. Retrieved from http://download.acca.it/Files/Relazione-Tecnica-Nuova-Sede ACCA.pdf. Accessed February 2018.
23. Mocerino, C. (2017). *Sistemi Intelligenti in Architettura. Seminario di Studi, Università di Salerno*. Rome: Gangemi Editore S.p.A. International. ISBN:978-884923548-7.
24. Market data: Energy efficient buildings – Europe Energy efficient commercial building technologies: Market analysis and forecasts for Western and Eastern Europe. Navigant research – Release Date: 3Q 2017. Retrieved from www.navigantresearch.com. Accessed January 2018.
25. Mocerino, C. (2017). Interoperable process: efficient systems in new constructive and product performances. *Journal of Civil Engineering and Architecture, 11*, ISSN 1934-7359.

EU-Financed LIFE-Diademe Project: Additional Energy Savings in Street Lighting by Means of IoT Sensors—A Case Study in Italy

Paolo Di Lecce, Andrea Mancinelli, Marco Trentini, Giuseppe Rossi, and Marco Frascarolo

Abstract Thanks to EU program LIFE, an innovative approach has been designed and soon will be completed in the city of ROME, within EUR district, with the project LIFE-Diademe. Today, the IoT technology (Internet of Things) makes easy to install, on each lighting pole, low-cost sensors, able to detect luminance, traffic flow and weather conditions. All these parameters can be measured in a more accurate way and, above all, in a wide urban area. Within the LIFE-Diademe project, 110 devices have been installed on 110 lighting poles and 890 more will be connected by the end of 2018 to measure, in a selected area, relevant parameters for Adaptive Lighting. To obtain a wide sampling of typical road lighting situation, the testing is considering urban contests representing different type of traffic: residential, offices, shops, Public Administration, University, etc.

On-site expert systems are analysing streets data and, thanks to the three basic evaluated parameters, they are adapting street lighting levels in real time mode: measurement and dimming time is being executed every minute.

First data about behaviour of the system are showing an approximate energy saving of about 30% compared to pre-programmed dimming cycles, and 50% compared to no dimming. These data are comparable to other Adaptive Lighting installations—designed according to standards—where the most significant result represents that in most of the urban roads, for 90% of the time, traffic flow is lower than 10% of nominal road capacity.

P. Di Lecce (✉) · A. Mancinelli · M. Trentini
Reverberi Enetec, Castelnovo nè Monti, Italy
e-mail: paolo.dilecce@reverberi.it; Andrea.mancinelli@reverberi.it; Marco.trentini@reverberi.it

G. Rossi
INRIM, Torino, Italy
e-mail: g.rossi@inrim.it

M. Frascarolo
Roma Tre University, Rome, Italy
e-mail: marco.frascarolo@uniroma3.it

P. Bertoldi (ed.), *Improving Energy Efficiency in Commercial Buildings and Smart Communities*, Springer Proceedings in Energy,
https://doi.org/10.1007/978-3-030-31459-0_7

Thanks to new IoT concepts, data about air quality, noise and post inclination are also being collected from each lighting point.

The LIFE-Diademe project experience will run for 1 year, to collect a reasonable set of data. After this period, a new lighting measurement campaign will be performed and, consequently, a Life Cycle Assessment (LCA) and a Life Cycle Cost Analysis (LCCA) are being carried out, in order to assess results, in terms of energy saving, safety, waste reduction, and, finally, sustainability.

1 What Is Dynamic Lighting

This case study is the update of previously published papers, which were presented the topic in an earlier stage of development (Di Lecce et al. *Outdoor adaptive lighting in the new UNI 11248 Italian standard and result of experience*. In Proceedings of the Lux Europa 2017 Conference Ljubljana Slovenia, September 18–20, 2017; Di Lecce et al. *Adaptive lighting in motorized traffic road: Real installations show that IOT technologies can support the correct use of standards*. CIE 2018, Smart Lighting Conference Taipei, April 24–28, 2018).

Relevant Topics
1. Lighting, appliances and equipment
13. Sustainable and smart communities, districts and cities.

1.1 *Dimming in the Last Two Decades*

The need of dimming, when late at night, was coming out in the 1970s, when the first energy crisis were obliging both citizens and governments to think and act towards energy conservation. The common feeling was that "all that light at night time, when nobody is there" would be a waste. During these years, in Italy, many municipalities decided to turn off one lamp out of two at midnight: a very simple technology, a good energy saving, but a much lower safety, considering that uniformity was altered and often lower than the minimum standard requirements.

In the late 1990s, a study from The Netherland motorways (DYNO) was carried out, to understand how to correlate traffic and weather conditions to safety and energy conservation. The result was detailing that during low traffic hours, lamps should be dimmed down to 20% (−80%), but only when meteorological conditions are fair, while in case of high traffic and bad weather, the lighting level should be double when compared to standards in force. The analysis of the behaviour of the vehicles over a prolonged period did not show any increase in critical situations, during reduced luminance time.

While CIE 115 was starting to consider dimming for energy conservation, in Italy the standard UNI 10439 introduced for the first time in Europe a very simple criteria to dim: it is allowed to downgrade of one lighting class, if the traffic volume is lower than 50% of the nominal value, and two lighting class whenever the traffic is reduced more than 75%.

Anyway, the idea was to give simple rules to allow energy conservations when peculiar conditions were met, so that no one should invent and follow his own specific approach to street light dimming.

2 LIFE: Diademe Project

In order to support the new technological step forward a full and real Adaptive Lighting, European Union program LIFE decided to finance an Adaptive Street Lighting demonstrative project.

In October 2016 a contract was signed between the Diademe consortium and LIFE program management, in order to demonstrate the potential of new Adaptive Lighting concepts, by using as reference starting point both CEN EN 13201 and UNI 11248 standards. This was a key point for the approval by the European Commission: the respect of standards gives the scientific framework to the demonstrative project. Results of energy saving, not linked to standards in force, can deliver poor or fantasist results, because the benchmark is not represented by energy saving itself, but energy saving in the framework of visual comfort and safety of vehicle drivers.

Project started in October 2016. In 2017 studies about new concept and installation constraints were defined and, the first 110 Diademe devices were produced.

The selected testing area within the well-known EUR district of Rome (Fig. 1) was really appreciated by the EU commission. Capitals have generally been

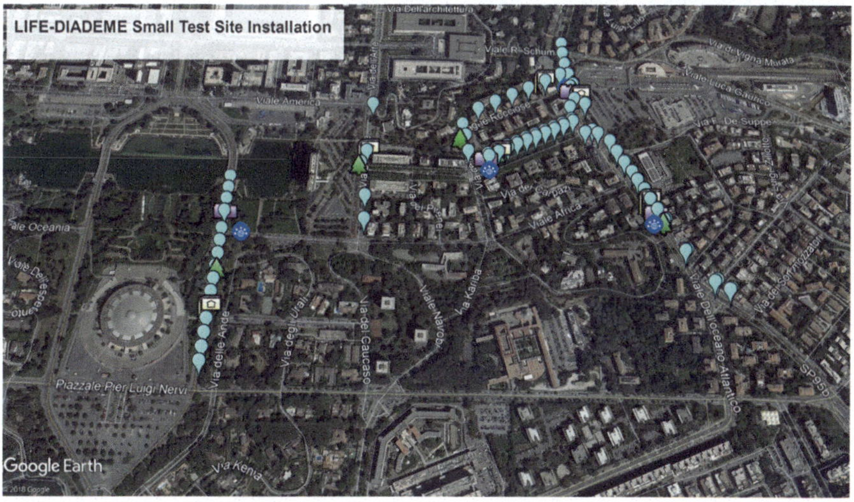

Fig. 1 Map of installation of first 110 devices

reluctant to adopt energy saving policies in street lighting; this is mainly due to the variability of conditions: critical traffic volumes at statistically unpredictable hours of the night, happenings, meetings, etc. EUR area in Rome is very varied and allows to record a wide sampling of typical road lighting conditions, because the tests considers urban contexts representing different traffic profiles: residential areas, presence of business and Public Administration offices, shopping malls, University, etc.

In early 2018, the available 110 devices were installed on the selected and measured roads.

The Diademe set of sensors (a device), installed in every pole of the road under analysis, includes the following:

(a) Luminance and traffic sensors, needed to run FAI adaptive lighting.
(b) Noise level.
(c) Vibration.
(d) Air quality.

Italian Standard UNI11248 has identified two ways to run Adaptive Lighting: TAI (Traffic Adaptive Installations), where only traffic flow is measured and FAI (Full Adaptive Installations), in which, besides traffic flow, luminance and weather conditions must be measured. FAI allows a reduction of three lighting categories, therefore a larger energy saving, while TAI only two.

Diademe project runs a FAI installation, in order to maximize the energy saving and the safety of road illuminance.

To evaluate weather conditions, since they normally do not vary for every lighting pole, five LTM sensors have been used. LTM sensor (Fig. 2) is a camera-based system, able to evaluate the three parameters relevant to UNI 11248, featuring state-of-the-art techniques of computer vision and image processing.

LTM is able to count the vehicles running across a virtual line, with an accuracy of about 10%. Vehicles are counted per each lane and per each carriageway, since UNI 11248 standard specifies that the traffic flow should be measured within the lane where the highest traffic volume is present.

Independently from the traffic volume evaluation, every minute, LTM takes a picture of the road, and calculates the average luminance of the area under analysis.

Fig. 2 LTM sensor and image analysis: measurement of Luminance, traffic flow and weather conditions

Image analysis is useful to determine as well if road is wet (uniformity analysis), if fog is present (condition where the dimming is being disabled), or if there is snow on its surface.

In order to be interesting as investment for the communities, PAY BACK for the investment should not exceed 5 years, therefore even if the energy saving could be quite interesting, the increment of installation cost due to Diademe should be quite low. Diademe is therefore verifying a new design approach, installing low cost IoT sensors, guaranteeing a full road coverage and granting a reasonable investment payback (less than 5 years).

Low-cost sensors are relatively available on the market, but the price paid is rising with the guaranteed accuracy. Traffic and luminance could be estimated, but in order to reach 5% accuracy in average luminance and 10% in traffic volume, many data coming from low accuracy sensors have to be correlated with high accuracy information calculated by the LTM. For Diademe project one LTM per 30 lighting poles has been installed, but preliminary estimation says that, if the correlation is fast and effective, it is possible to install one LTM every 300 lighting poles.

In order to provide a higher added-value to Diademe project, different sensor types have been installed (Fig. 3): noise sensor for every luminaire, in order to map noise level, according to EU directives; air quality sensors (one every 20 lighting poles), in order get info about gases in the atmosphere; vibration and inclination sensors, in order to increase safety by avoiding unexpected luminance distribution and pole instability problems.

Fig. 3 Example of LIFE Diademe installation on lighting posts

3 Diademe Results

First data coming from 110 sensors are showing that a good 30% of energy saving, compared to pre-programmed dimming cycles, could be achieved, and this value rise at 50% of energy saving if no dimming is in force (Fig. 4).

Looking at the detail, we are noticing how the traffic volume, during night hours, is much lower than during daytime. Considering that each single road has a specific capacity relevant to traffic during the day (traffic density), we can then conclude that most of the time during the night, the road is over-lit.

A deeper analysis is showing, in fact, that in many of the sampled roads, the traffic is lower than 12% of its nominal value for more than 90% of the night time. This evidence is explaining why the targets defined in the LIFE-Diademe project can be achieved (Fig. 5). Using LED light sources, and UNI 11248 prescription, lighting can be downgraded by three lighting classes, that is with a 70–80% of dimming during most of the night, ensuring that driver safety is not be compromised.

LIFE Diademe project is installing a total of 1000 Diademe devices within EUR district by the end of 2018.

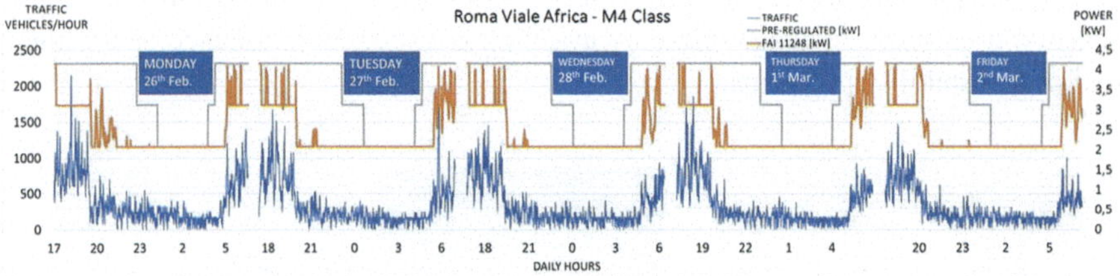

Fig. 4 Traffic and power data acquired during March 2018 on Viale Africa

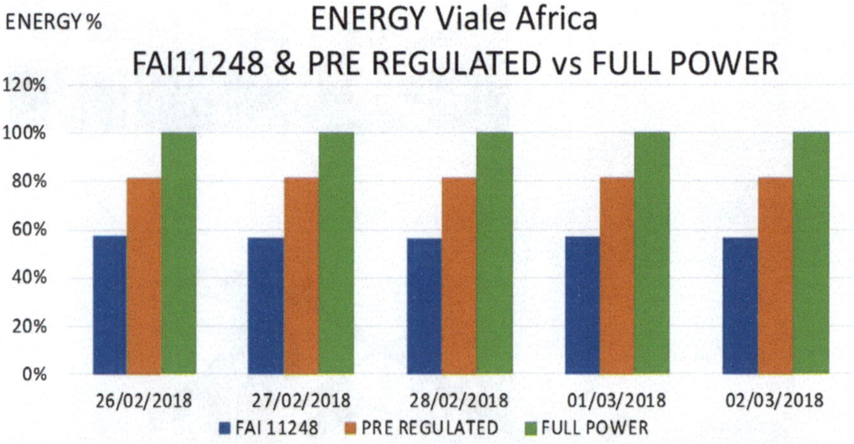

Fig. 5 Energy saving (in %) in a week of the Viale Africa—March 2018

4 Conclusions

The LIFE-Diademe project experience will then run for 1 year after the complete installation of all devices. This is allowing to collect a reasonable set of data. After this period, a new lighting measurement campaign is being performed and, consequently, a Life Cycle Assessment (LCA) and a Life Cycle Cost Analysis (LCCA) analysis will be carried out, in order to assess results, in terms of energy saving, safety, electronic waste reduction, and, finally, sustainability.

EU program LIFE considers very important LCA and LCCA analysis. Only a careful and complete analysis of energy spent to produce the system and to waste it at the end of its life cycle makes such a system useful. Since dimming is reducing running temperature of LED source, we expect to increase lamp life by at least 10%. The reduction of Waste Electrical & Electronic Equipment (WEEE) is another important target of LIFE-Diademe project.

Data about air quality are being processed by a special self-learning Big Data algorithm. Low precision sensors data need to be adjusted during their life time according to weather parameters, and only a significant number of sensor in a small area (e.g. in EUR 50 of these sensors are being installed), with a good data elaboration software, can give reliable and accurate information.

Acknowledgement We would like to thank EU program LIFE, which financed 60% of DIADEME LIFE15 CCM/IT/000110, allowing us to install such an interesting and innovative system.

Further Reading

1. Ministry of Transport, Transport Research Center, The Netherlands—*Dynamic public lighting—cover report—Marzo 1999.*
2. CEN. (2014). *CEN TR13201-1:2014. Road lighting—Part 1: Guidelines on selection of lighting classes*. Brussels: CEN.
3. CEN. (2015). *EN 13201-4:2015. Road lighting—Part 4: Methods of measuring lighting performance*. Brussels: CEN.
4. CEN. (2015). *EN 13201-5:2015. Road lighting—Part 5: Energy performance indicators*. Brussels: CEN.
5. CEN. (2015). *EN 13201-5:2015. Road lighting—Part 2: Performance requirements*. Brussels: CEN.
6. CEN. (2015). *EN 13201-5:2015. Road lighting—Part 3: Calculation of performance*. Brussels: CEN.
7. CIE. (2001). *CIE 144:2001. Road surface and road marking reflection characteristic*. Vienna: CEN.
8. CIE. (2003). *CIE 154:2003. The maintenance of outdoor lighting system*. Vienna: CEN.
9. CIE. (2010). *CIE 115:2010. 2nd editon. Lighting of roads for motor and pedestrian traffic*. Vienna: CEN.
10. ISO/CIE. (2014). *ISO/CIE 19476:2014. Characterization of the performance of illuminance meters and luminance meters. Joint ISO/CIE international standard*. Geneva: ISO.
11. UNI. (2011). *UNI 11431:2011. Luce e illuminazione—Applicazione in ambito stradale dei dispositivi regolatori di flusso luminoso*. Milano: UNI.

12. UNI. (2016). *UNI 11248:2016. Illuminazione stradale—Selezione delle categorie*. Milano: UNI.
13. Di Lecce, P. (2016). *Illuminazione adattiva: Italia sempre un passo avanti* (pp. 32–33). Luglio Agosto: U&C (Unificazione & Certificazione, rivista ufficiale di UNI). n.7.
14. Rossi, G., Iacomussi, P., Mancinelli, A., & Di Lecce, P. (2015). Adaptive systems in road lighting installations. [INRIM, Torino, ITALY], 2. REVERBERI Enetec, Castelnuovo ne' Monti (RE), Italy. *Light & Engineering, 23*, 4.
15. Rossi, G., Iacomussi, P., Mancinelli, A., & Di Lecce, P. (2015). *Adaptive systems in road lighting installations. [INRIM, Torino, Italy], 2. REVERBERI Enetec, Castelnuovo ne' Monti (RE), Italy. CIE 216*. In Proceedings of the 28th Session of the CIE, 28 June–4 July 2015, Manchester, UK.
16. Di Lecce, P., et al. Lux Europa Lubjiana September 2017. Outdoor adaptive lighting in the new UNI 11248 italian standard and result of experience (OM09).
17. Di Lecce, P., Mazzocchi, A., Mancinelli, A., & Rossi, G. (2017). Illuminazione adattiva: l'Italia fa sul serio. *LUCE, 322*, 90–94.
18. Di Lecce, P., Mazzocchi, A., Mancinelli, A., Rossi, G., & Frascarolo M. (2018). Adaptive lighting in motorized traffic road: *Real installations show that IOT technologies can support the correct use of standards*. CIE Topical Conference on Smart Lighting in Taipei, Chinese Taipei, April 26–27, 2018.

Building Integrated Photovoltaic Systems as a Sustainable Option for Retrofitting of Office Buildings in South East Europe

Anna Serasidou and Georgios Martinopoulos

Abstract The need for rational energy consumption and measured use of resources dictates a new approach to designing, constructing, and renovating existing buildings. This paper focuses on one of the main energy consumers within the built environment, office buildings.

In order for office buildings to comply with the targets set for 2020 by the Energy Performance of Buildings Directive, extended refurbishment of the existing building stock is required, combined with utilizing renewable energy technologies. Although there are various strategies available for renewable energy generation in urban environments, facade BIPV integration offers a great potential of generating electricity, despite the limited roof space of multistory buildings.

The case of buildings in Southeast Europe is of special importance, as due to the prevailing climatic conditions, cooling loads are usually higher than heating loads, making retrofitting a more complex problem than simply increasing the insulation levels.

For the scope of this paper, the facade redesigning of a typical nine-story office building in Greece is examined as a sustainable option towards transforming it into a nearly Zero Energy Building (nZEB). In order to achieve greater energy performance, an energy simulation model is developed in EnergyPlus and TRNSYS, to calculate the energy savings and electricity production through the proposed retrofitting options. The BIPV systems are estimated to produce electricity that covers approximately 50% of the building's total annual energy demand and upgrade its aesthetics and architectural form. Moreover, various orientation scenarios are evaluated, to better understand the behavior and retrofitting potential of offices scattered throughout the urban environment of Southeast Europe.

A. Serasidou (✉) · G. Martinopoulos
School of Science and Technology, International Hellenic University, Thessaloniki, Greece
e-mail: a.serasidou@ihu.edu.gr

P. Bertoldi (ed.), *Improving Energy Efficiency in Commercial Buildings and Smart Communities*, Springer Proceedings in Energy,
https://doi.org/10.1007/978-3-030-31459-0_8

1 Introduction

The built environment occupies a significant place in every aspect of everyday life. Its environmental impact is substantial, and its energy performance is so poor that buildings are placed among the major CO_2 emitters in Europe. More specifically, the building sector is responsible for approximately 40% of the EU's total final energy consumption and CO_2 emissions [1, 2].

Since the beginning of the twenty-first century there has been a change in the public's perception of these environmental issues. The "Energy Performance of Buildings Directive" (EPBD) and its recast [1] are nowadays the main legislative tools in the EU that provide a holistic approach towards energy efficiency in the buildings sector. They have set targets for 20% reduction of energy consumption and greenhouse gas emissions, and 20% increase in the share of renewables by 2020 [1]. Furthermore, they have introduced the requirement for better energy performance not only for new buildings but also for the vast majority of existing ones, that are in an acute need of refurbishment [3]. The need for rational energy consumption and measured use of natural resources dictates a new way of thinking about the design, construction, and renovation of buildings. This paper focuses on one of the main energy consumers within the built environment, office buildings.

Commercial buildings are, nowadays, an integral part of the building stock of urban areas in most of the developed countries all over the world. Based on data collected across the EU, it is estimated that the average energy consumption in the nonresidential sector reaches 250 kWh/m^2, for all end uses [4, 5]. This consumption is about 40% higher than the respective value for the residential sector [4, 5]. Undoubtedly, variations exist from one country to another and from building type to building type, but office buildings, while not being as energy intensive as hospitals or hotels, represent more than 26% of the energy use in the nonresidential sector [4, 5].

The amount of energy used in modern office buildings, is a complex issue. Most of the energy in the nonresidential sector and, more specifically in office buildings, is used for heating, lighting, computing, and hot water [6, 7]. Thus, office buildings have the tendency to be internal load-dominated. The case of buildings in the area of South East Europe is of special importance, as due to the prevailing climatic conditions, cooling loads are in some case higher than heating loads, making retrofitting of the buildings a more complex solution that simply increasing the level of insulation [8]. In order for this type of buildings to comply with the EPBD and the 2020 targets, there is a great need for refurbishment and extended use of renewable energy technologies for covering their energy demand.

It is apparent that in order to reduce energy use in the building sector it is not enough to make only new buildings more energy efficient or even nZEBs, as the current, almost nonexistent renovation rate of about 1.2% annually [9] should be greatly increased. Retrofitting of existing buildings on the other hand is a viable solution as it presents multiple benefits. First of all, there are economic advantages associated with increased energy retrofitting rates. The stimulation of the building industry and the creation of new jobs in sectors related to building renovation are at the forefront. Furthermore, organized retrofitting efforts result in increased asset values and a general acceleration of national economic growth [10]. In order for the total lifetime costs of a retrofitted building to be considered, cost optimal methodology is a key driver for the evaluation of the operational, maintenance, disposal and energy saving costs apart from the initial investment cost [11].

The environmental benefit from the retrofitting of office buildings is the reduction of greenhouse gas emissions, and especially CO_2 and thus, mitigation of climate change impacts taking into consideration the whole life cycle of the retrofitted building; that is taking into consideration the emissions generated during the production and use of the materials and energy systems employed in the retrofit. Various studies have proved that such retrofitting solutions have a positive life cycle impact, mostly due to the large amount of emissions attributed to the use of fossil fuels [12]. Moreover, the benefits for the society are primarily the limitation of energy poverty and social exclusion. Energy retrofitting could decrease the heating and cooling running costs and contribute to improving the quality of life and thermal comfort of occupants in the office premises.

For the scope of this paper, the facade redesigning of a typical nine-story office building in Greece is examined as a sustainable option towards transforming it into a nearly Zero Energy Building. In order to achieve greater energy performance, an energy simulation model is developed in EnergyPlus and TRNSYS, to calculate the energy savings and electricity production through the proposed retrofitting options. Moreover, various orientation scenarios are evaluated, to better understand the behavior and retrofitting potential of offices scattered throughout the urban environment of Southeast Europe.

2 Building Retrofitting Options for Achieving nZEB Goals

Due to the great diversity of the European building stock, there cannot be one simple strategy to tackle the various energy challenges of the urban environment. Numerous studies are carried out in order to evaluate the parameters that affect a building's energy performance and possible measures to improve it. The retrofitting proposals most commonly encompass high performance building envelope, lighting system, and HVAC systems, combined with the use of RES for on-site energy production [13].

One of the first approaches in literature for energy retrofitting was the study and review of facade design and construction technologies. Case studies examine building envelopes with improvements to their thermal characteristics, cool materials [14], replacement of windows, increased airtightness, use of solar shading systems [15], and active cooling measures, such as natural and night ventilation [16].

Furthermore, nowadays, that there is great interest in energy efficient and cost-effective building system technology for nZEBs, the goal of recent case studies is to optimize the utilization of RES, especially in urban areas. These technologies can act as an alternative for reducing fossil fuel dependency and thus, carbon emissions, and promote the construction of "greener" buildings, especially for the nonresidential sectors [17]. Most researchers focus, on the incorporation of BIPV technologies. Solar power is the logical winner among the existing RES technologies available, as rooftops and facades offer ideal fields for solar energy exploitation. PVs are a suitable choice as the vast majority of both residential [18] and nonresidential, and more specifically office buildings, operate during daytime, when energy production by solar radiation is at its maximum. Furthermore, in facade renovation scenarios BIPVs are a great choice, since they can substitute common cladding materials, offering smaller amortization periods and giving a differentiated architectural character in the buildings [19].

On the other hand, measures regarding HVAC systems can combine improvements on heating, cooling, and ventilation systems. Common heating system improvements suggested in literature include the use of more efficient boiler systems, heat recovery, and improvements on the distribution systems [20]. Another noteworthy approach, nowadays, is the combination of both energy efficiency and on-site energy production technologies with the use of heat pumps. Heat pumps are promising systems, due to their high-performance values with low loads, enabling low supply temperature and operating with the electricity produced by BIPV on-site. Therefore, heat pumps have been thoroughly investigated by the International Energy Agency in the Annex 40 of the Heat Pumping Technologies Program (HPT), both in simulation case-studies and field monitoring across different countries, towards achieving the nZEB goals [21].

Of course it should be noted that, retrofitting a building to reduce its energy consumption and, subsequently, the energy costs should always be studied in a way that improves the aesthetics and increase the value of the asset, providing at the same time better living or working conditions for the users [5].

3 Case Study: High-Rise Office Building in Greece

The Greek building stock, according to the Hellenic Statistical Authority, comprises of 4,105,637 buildings, the vast majority of which—approximately 80%—are single floor residences [22]. Mixed use high-rise buildings in the Municipality of Athens and Thessaloniki with dominating office usage correspond to approximately 8% of the total building stock [23].

The first legislation in Greece regarding building energy efficiency was the Thermal Insulation Regulation of buildings, introduced in 1979 [24]. The majority of buildings, almost 60% [25], was however, constructed before 1980, resulting in them having a very poor energy performance. Additionally, the implementation of the EPBD was rather delayed, as it became part of the national legislation in the summer of 2010 [26]. Therefore, the vast majority of buildings are uninsulated, experience high energy losses through their envelopes, and lack in the utilization of state of the art HVAC equipment.

According to data gathered by the Ministry of Environment, Energy and Climate Change through the Energy Performance Certificates (EPC), approximately 40% of the buildings examined are in the energy efficiency bands F and G, while only 3% of the audited buildings are in bands A+ to B+ [27]. The average energy demand of a building that was built before 1980 is almost 162 kWh/m^2/year [28]. These buildings have, however, nowadays a significant potential for large scale retrofitting, in order to achieve reduced energy consumption and better living conditions for the occupants. In order for the renovation programs to be effective though, in terms of both energy efficiency and economic viability, a deep understanding and recording of their existing condition is crucial.

Typical construction material, in almost 65% of the cases, is reinforced concrete. The building envelopes consist of brick masonries with double brick cavity walls and extended glazing surfaces, where office use is dominant. Another noteworthy fact is that 40.4% of the high-rise urban buildings have flat roofs [22], with great solar and wind access. Although this allows for higher energy consumption and poorer comfort conditions for top floors throughout the year, it creates ideal spaces for renewable energy systems integration.

The selected case study building possesses a variety of typical characteristics by being constructed with typical materials, at a period when most of the buildings throughout the city were built, utilizing typical energy sources, that is, natural gas and electricity, and operating at a typical 8 h work schedule. It is located in the dense urban center of Thessaloniki and it was constructed in 1968. It is a semi-detached building with a net surface of 1488 m^2 and a volume of 4300 m^3. It consists of a basement with storage spaces, a ground floor with a reception area and storage spaces, eight upper levels with offices, and a ninth mechanical and storage floor. It is the administration building of a constructional company, housing 40 employees daily. As the floor plan shows (Fig. 1), the shape of the building is an irregular pentagon, elongated along the N-S axis.

It occupies a corner plot and is adjacent to two other office buildings on its north and northeast sides, while being detached in the three remaining facades. Each level is 3 m-high, making the total height of the building 30 m. The surrounding buildings are mostly of the same height, except for the building opposite the northeast side, which is only five floors tall, enabling better insolation in this side. The total heated surface of the building is 1000 m^2, split among the first to eighth floor, as presented in Table 1.

The building was constructed based on the, then current, building standards and regulation, that is, the National General Building Regulation of 1955 [29]. The materials used for the construction of the building were the typical ones used in Greece for high-rise buildings during its construction era. The bearing structure is of reinforced concrete and the outer walls are double brick masonries with no insulation. The flat roof slabs have no insulation either. The cladding of the facade consists of aluminum panels with a total thickness of 3 mm which are supported by a metallic structure allowing a gap of approximately 5 cm between the external wall and the cladding. The building is equipped with sliding double glazed aluminum windows with no thermal or sound insulation. Windows cover approximately 65% of the building's facade. The building element characteristics are summarized in Table 2.

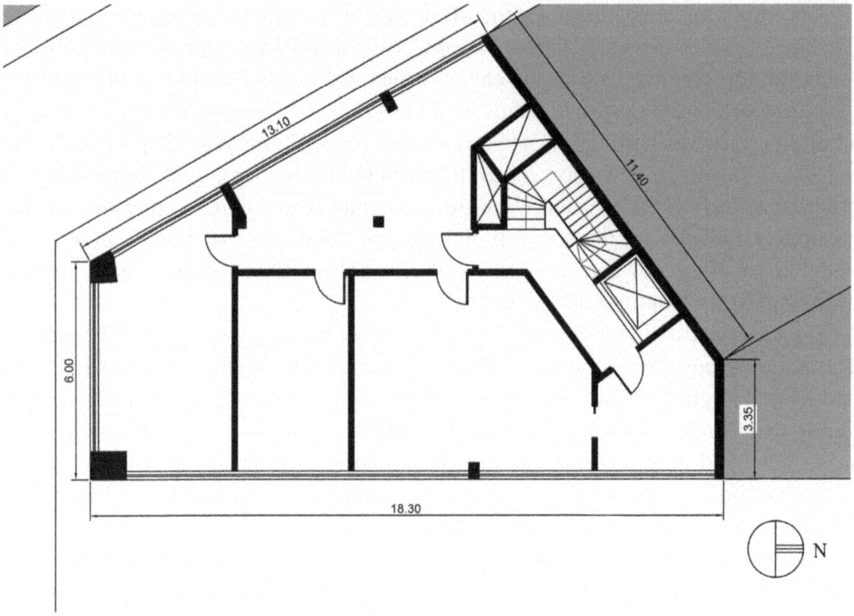

Fig. 1 Floor plan of typical levels (1–8)

Table 1 Breakdown of the building's surfaces

	Heated surface (m^2)	Unheated surface (m^2)
Basement	–	170
Ground floor	–	142
Typical floors (1st to 8th)	125	17
9th Floor	–	40
Total	1000	488

Table 2 Building elements characteristics

Building elements	*U*-values (W/m^2K)
Double brick wall with metal sheet outer layer (no insulation)	1.049
Concrete slab with tiled floor with no insulation (ground floor)	2.5
Concrete slab with laminate flooring with no insulation (offices)	2.10
Flat roof	3.3
Doors/windows	3.1

The heating of the office building is achieved with a centralized, natural-gas-fired boiler, with a nominal capacity of 349 kW. The boiler is coupled to radiators and the heating system is connected to a central thermostat for its optimal operation.

For cooling purposes split-type air-conditioning units are utilized, mounted on the walls of each office. Most of the units have a cooling capacity of 5 kW and an EER 3.23. These units are manually operated by the occupants of each office, thus, there are no standard set points. There is no mechanical ventilation system in the office building, as per usual practice in Greek buildings, as the existing regulations in Greece do not impose the use of mechanical ventilation [26], contrarily to other EU members. Natural ventilation through the building's windows enables the required fresh external air to enter the building.

The lighting of the offices and the hallways is achieved by fluorescent ceiling lights. All lighting fixtures are manually operated since there are no automations installed depending on occupancy or illuminance. Regarding the office equipment, the building is moderately computerized. The mean installed power of the office equipment and appliances is simulated as 17 W/m^2 for the heated zones. The respective power for the lighting fixtures is 12 W/m^2 for the heated zones and 10 W/m^2 for the unconditioned zones.

Based on historical data collected from the building's electricity bills from the past 5 years, from 2012 until 2016, the average annual electricity consumption is 92,310 kWh. The declining rate also reflects an effort to reduce energy costs, which conforms to a general tendency in both Greece and the EU, during the previous years, for reduction in electricity consumption due to the economic crisis [30].

4 Simulation of the Baseline Building

Selecting the correct and suitable retrofitting measures for every building can prove to be a very challenging task which requires a vast amount of information to be processed. Therefore, building energy simulation software are, nowadays, an indispensable tool for designers, engineers and manufacturers, in order to decide upon and complete a retrofitting project, and deal with any uncertainty or risk that may arise during their assessment. A comprehensive energy audit needs to be made, thus a systematic procedure to record and assess the existing condition and the energy consumption profile of the building, and identify and quantify cost-effective energy saving opportunities [31].

Subsequently, an energy simulation model is developed, with the widely used simulation software EnergyPlus [32, 33] in order to evaluate the existing building's performance and afterwards estimate energy savings through the proposed design options. For the scope of this paper the simulation software EnergyPlus is selected. EnergyPlus performs sub-hourly calculations and integrates the load and system dynamic performance into the whole building energy balance calculations [34]. Heating and cooling loads are calculated by the software, on an hourly basis, by the heat balance method [35]. The geometry of the buildings is designed in SketchUp [36] with the use of the free extension Euclid [37].

The potential for electricity production through the BIPV systems is estimated with the use of System Advisor Model (SAM) [38] which is used worldwide by academics and professionals analyzing, designing, and installing RES [39–42]. SAM is based on the dynamic simulation engine of TRNSYS [39] and uses hourly climatic data in the form of a Typical Meteorological Year. By providing the software with the geometrical characteristics of the PV panels, orientation, efficiency, and cost data of the whole installation, the annual electricity production is calculated, along with the energy savings and the respective amortization period.

Moreover, in order to achieve a better understanding of the behavior and retrofitting potential of other office buildings in the city, various orientation scenarios for the reference building are evaluated through the software. The building is simulated for a period of a whole year, from 01/01 until 31/12 using the International Weather for Energy Calculations (IWEC) of Thessaloniki with hourly values [34]. The zoning of the building is done according to the use of each level. The model consists of 11 zones. Each heated level has four offices (104 m^2), one reception area (38 m^2) and the staircase area (17 m^2). All the internal doors of the office level remain open throughout the day; thus, each level is considered to be one separate thermal zone, in order to better study their energy behavior. However, the staircase area is separated from the office zones in each level, as it is an unheated space and is simulated as a separate unconditioned Zone. Ground floor (Zone_0) and the ninth floor (Zone_9) are also unconditioned spaces.

The Zones, as the actual building itself, are exposed to the outside air on the East, South and Southwest sides and are adiabatically in contact with the adjacent buildings on the North and Northwest sides. The first floor (Zone_1) is adjacent on its lower side to the unheated ground floor and eighth floor (Zone_8) to the external air. All the other intermediate levels (Zone_2 to Zone_7) have almost identical characteristics and are adiabatically connected with the other heated zones through their floor and ceiling.

The occupancy schedule of the office building is from Monday to Friday, 07:00–19:00, during week days. The office building is simulated as closed during weekends and the official Greek holidays. The people occupying the office building on a daily basis are evenly distributed among the heated levels. Therefore, five people are occupying each of the heated Zones during the operating hours in the simulation for a total of forty. The lighting fixtures and the equipment are considered switched on during occupancy hours. Their operating profile is presented in Figs. 2 and 3 respectively.

Since the building is located in a dense city center, all the neighboring buildings are considered as shading objects in the simulation for a realistic result, taking into account their respective heights and the width of the surrounding roads and pavements (Fig. 4).

Regarding the HVAC systems, an ideal system is simulated, considering a heating period from 01/01 until 30/04 and 20/10 until 31/12 and a cooling period from 01/05 until 19/10. The set temperature points in the typical office level zones are

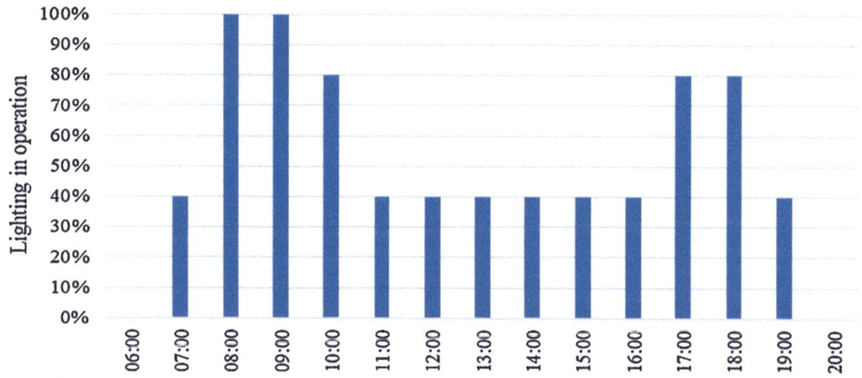

Fig. 2 Lighting operation profile

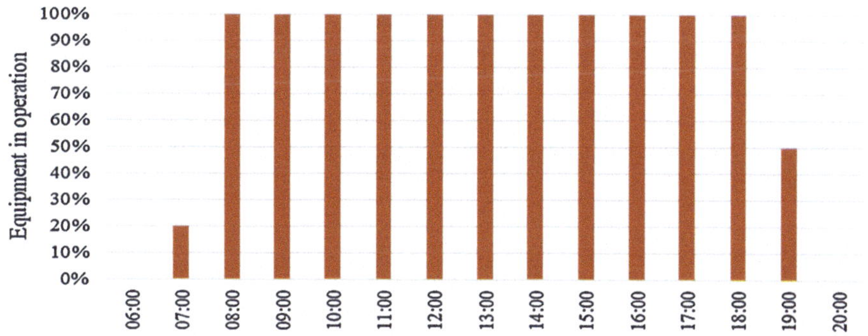

Fig. 3 Equipment operation profile

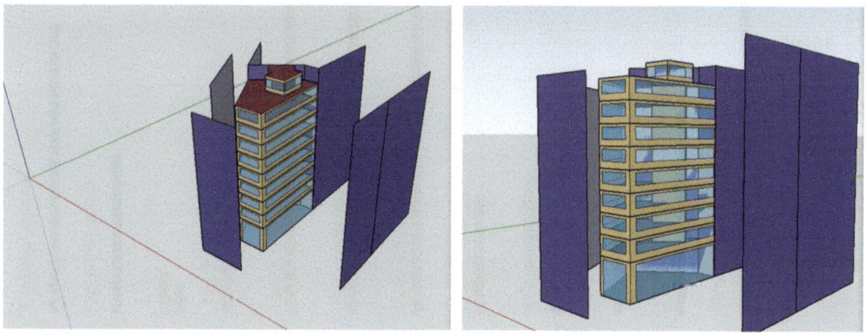

Fig. 4 Building model with the constructive elements of the envelope and the surrounding shading surfaces (SketchUp—Euclid)

26 °C for cooling (with a tolerance of ±1 °C) and 22 °C for heating (with a tolerance of ±1 °C) [43]. No thermostats are set in the unheated zones.

The simulation provides a total annual energy demand of 159,074 kWh for the office building. Therefore, the energy demand per conditioned building area is 159.07 kWh/m². The demand for heating is 46,282 kWh, for cooling 59,277 kWh and the respective value for lighting and equipment is 53,515 kWh, as presented in Fig. 5. The cooling demand is almost 25% higher than that for heating, as it is expected in an office building dominated by internal loads.

It can be observed that, as there is no insulation in the building, the envelope responds quickly to fluctuations of the outside air when the HVAC systems and the lighting and equipment are not in operation. Figures 6 and 7 display the mean air temperature of each zone at a daily basis, in comparison with the outside temperature, for the most energy intensive months, January and July. It can be observed, that during the heating period the office building maintains a mean temperature of approximately 19 °C during the week days (Fig. 6). The mean temperature during weekends follows the fluctuations of the outside weather, with Zone_8 being the most volatile and susceptible to temperature changes.

Similar results emerge during the cooling period as well. The mean temperature of the office building is approximately 29 °C during week days, while being prone to dramatic changes, depending on the outside temperature, during holidays and weekends (Fig. 7).

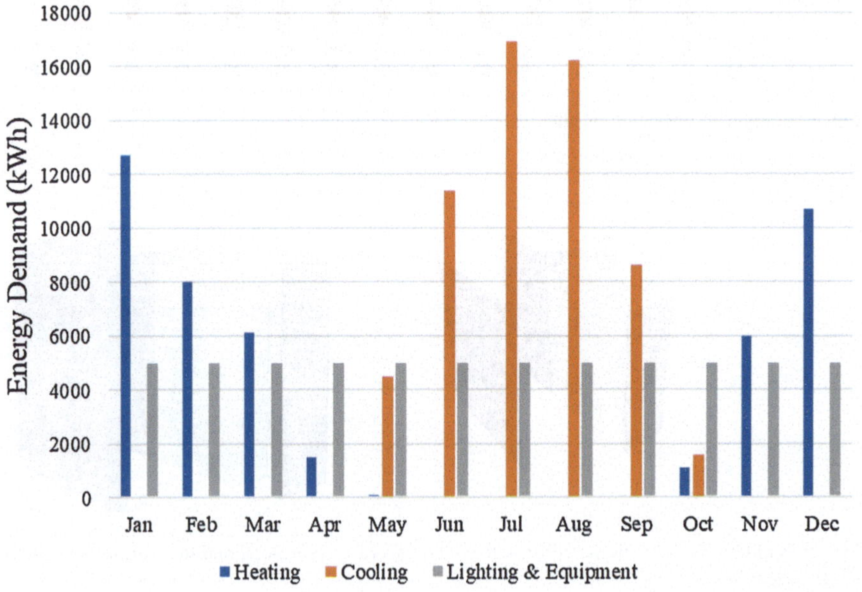

Fig. 5 Monthly energy demand in kWh calculated by EnergyPlus

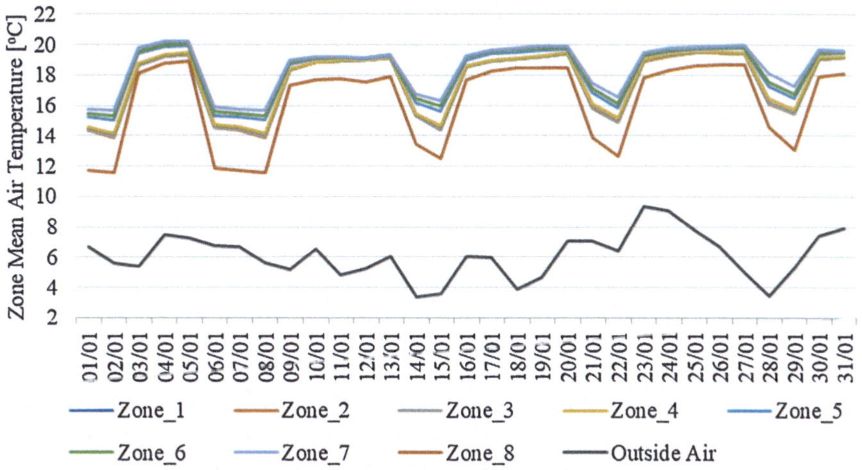

Fig. 6 Evolution of daily mean air temperature per Zone during January

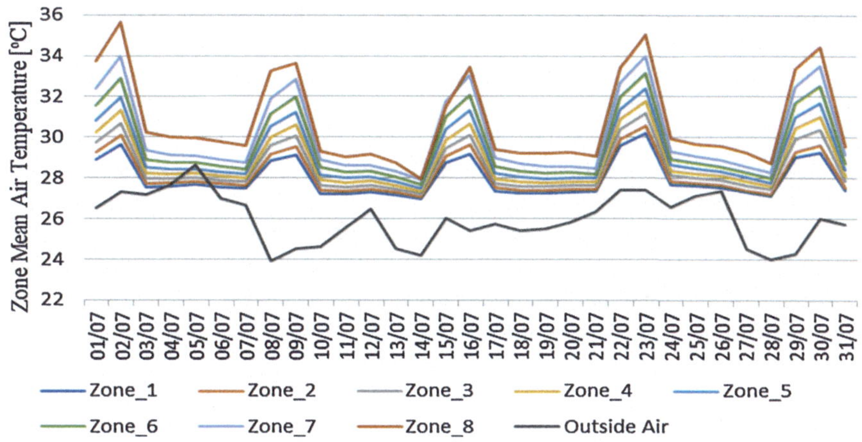

Fig. 7 Evolution of daily mean air temperature per Zone during July

5 Design Retrofitting Scenario Results

Following the thorough literature research and the evaluation of the existing condition of the office building, design proposals are investigated in order to improve the building's energy performance. The main goal is to cover most of its energy needs through RES and achieve energy savings.

The proposal consists of a facade retrofitting with the installation of Building Integrated Photovoltaics on the three detached facades of the building and its roof. The BIPV systems offer solutions that enhance the versatile function of the envelope.

With the technological development, BIPV systems do not only offer great potential for generating electricity but also upgrade the aesthetical and architectural form of a building as a whole [19].

The proposal involves the installation of the PVs on the opaque parts of the facade by substituting the existing composite aluminum cladding panels. Thus, the window area and the window to wall ratio of the building will remain the same for the scope of this scenario.

Since the efficiency of the panels is approximately inversely proportional to the cell temperature [44], the facade modules should be placed at a distance from the outer wall. High cell temperatures due to overheating could more easily occur for BIPVs compared to a free standing installation, because of the low wind cooling effect [44]. Therefore, it is important to provide an adequate air gap behind the modules to allow for air circulation and cooling. More specifically, in this scenario the air gap of the 5 cm behind the existing cladding is considered to remain on the back of the modules as well, facilitating better the heat rejection process and avoiding efficiency losses.

The PV modules are incorporated in the East, South and Southwest facades, and the two free parts of the flat roofs of the eighth and ninth floor. Each facade installation could incorporate eight horizontal bands of modules, one for each level, placed underneath the window area, as displayed in Fig. 8. They cover a total area of approximately 280 m^2—more specifically, 140 m^2 in the East, 38 m^2 in the South, and 102 m^2 in the Southwest facade.

Depending on the dimensions of each facade a total of 176 panels could be installed in the building, with the technical characteristics presented in Table 3, with a total maximum power of approximately 58,000 kWp. The East facade has a string of 88 modules spread on the eight floors, with a tilt of 90° and an azimuth of 90° (North being 0°). The South facade has a string of 24 modules with a tilt of 90° and an azimuth of 180° and the Southwest facade has respectively a string of 64 modules with a tilt of 90° and an azimuth of 240°.

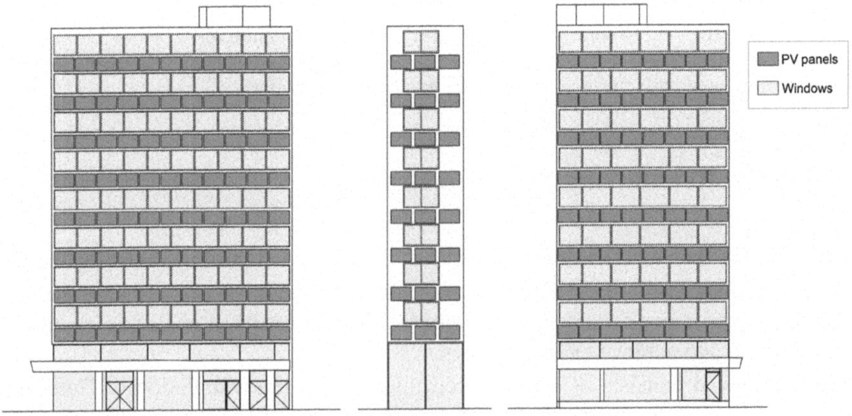

Fig. 8 Elevations incorporating the PV modules

Table 3 Technical characteristics of the PV panels

Maximum power (P_{max}) [V]	330
Max power voltage (V_{mpp}) [V]	33.7
Max power current (I_{mpp}) [A]	9.8
Open circuit voltage (V_{oc}) [V]	40.9
Closed circuit current (I_{sc}) [A]	10.45
Module efficiency [%]	19.3
Dimensions ($X \times Y \times Z$) [mm]	1000 × 1590 × 50
Operating temperature [°C]	−40 °C to +90 °C

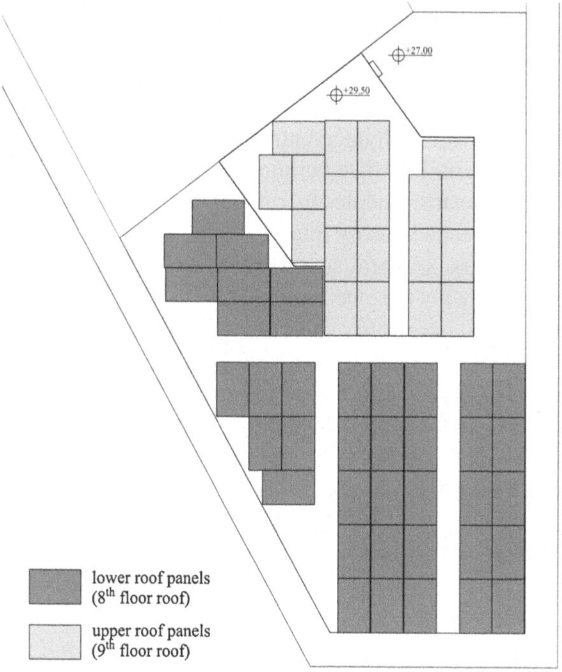

Fig. 9 Plan of the flat roofs with PV panels installed

The roof installation is split between the eighth and ninth floor roofs. The flat roof of the eighth floor is approximately 100 m^2 and the respective of the ninth floor is 40 m^2. Because of the limited space available, the panels are installed horizontally on the flat roof, with a tilt of 0° and an azimuth of 180°, in order to avoid any self-shading, and maximize the number of panels that can be installed. As Fig. 9 shows, 40 panels can be installed on the roof of the eighth floor and 19 more above the ninth floor. The proposed panels have a total maximum power of approximately 19,500 kWp.

According to the existing legal framework in Greece [45], the PV arrays should be connected to the utility grid via inverters and operate under the net-metering scheme. Since the modules are designed to be integrated in various orientations, the use of multiple string inverters was selected. The selected inverters have a maximum capacity of 4000 W and efficiency of about 95%. The arrays on the East facade require five inverters, while the ones on the South and Southwest facade two and three respectively. The upper roof arrays require two similar inverters and the lower ones three more, making a total requirement of 15 inverters for all the building installations.

5.1 Impact of Orientation on the nZEB Aspect of the Building

Since the building possesses various typical attributes of office buildings in Greece and Southeast Europe in general, a parametric analysis of these scenarios is conducted with respect to the building's orientation. Therefore, the energy simulation models are developed respectively, with two extra orientation scenarios which are examined and compared with the baseline case. Thus, valuable results are deduced for the effect such interventions could have, not only on the examined building but on other similar office buildings throughout the country. The first alternative orientation scenario emerges by rotating the existing, baseline building by 90° clockwise and the second one by rotating the building 45° clockwise (Fig. 10).

The first scenario serves in allowing the largest facade of the building to face absolutely towards the South, while the second one allows all three of the detached facades to have better solar access. By applying such rotations, the electricity yield of the PV installations is obviously altered. Minor deviations between the scenarios are also expected in the overall energy performance of the building, regarding the heating and cooling demands. In both cases all the other intervention parameters, such as the number of PV panels installed or the thickness of the insulation layers, and their respective cost analysis remain the same.

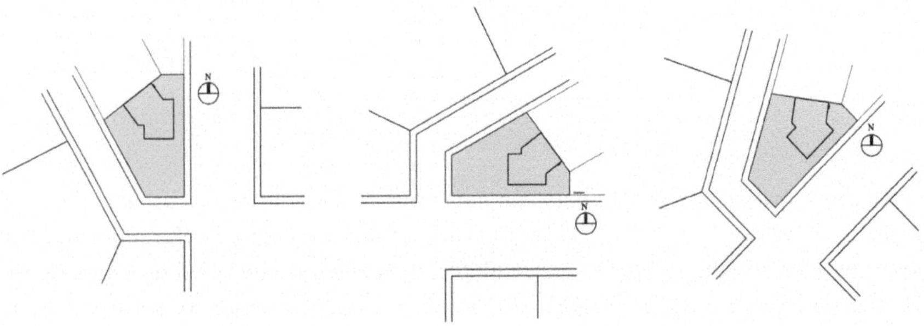

Fig. 10 Alternative orientation scenarios: (**a**) baseline, (**b**) 90°, and (**c**) 45°

6 Results

In order to achieve more realistic results a different simulation was run for each different system orientation. The results, presented in Fig. 11, reveal that the total amount of energy produced by the modules can be 68,764 kWh. This production could cover up to 81% of the total electricity needs of the building.

It is cost effective for an installation when the utility load and the renewable energy generated are well matched. The total cost for the BIPV installation on the facades and roofs is calculated at 120,000€, which corresponds to 1.55 €/Wdc.

The cost effectiveness of the proposal is determined with the Simple Payback Period Method [46]. The current electricity rates of the Public Power Corporation (P.P.C.) for companies is 0.17 €/kWh [47]. Based on this price, the annual cost savings could be up to 11,669€ leading to an amortization period of approximately 10 years.

The investigation of the various interventions applied in the selected building revealed some noteworthy results. Regarding the electricity production of the BIPV systems, small differences can be observed between the examined scenarios. The greatest production, of 69,189 kWh, is achieved for the 45° rotation scenario. However, the energy yield is only slightly improved compared to the initial scenario (Fig. 12). A bigger reduction can be observed in the load of the 90° rotation, reaching 64,932 kWh. Since, the photovoltaic panels incorporated into the opaque elements of the facades are vertically integrated, with a tilt of 90°, the South orientation is not ideal, whereas the East and West ones are better suited. This explains the better energy yield of initial and the 45° rotation scenarios.

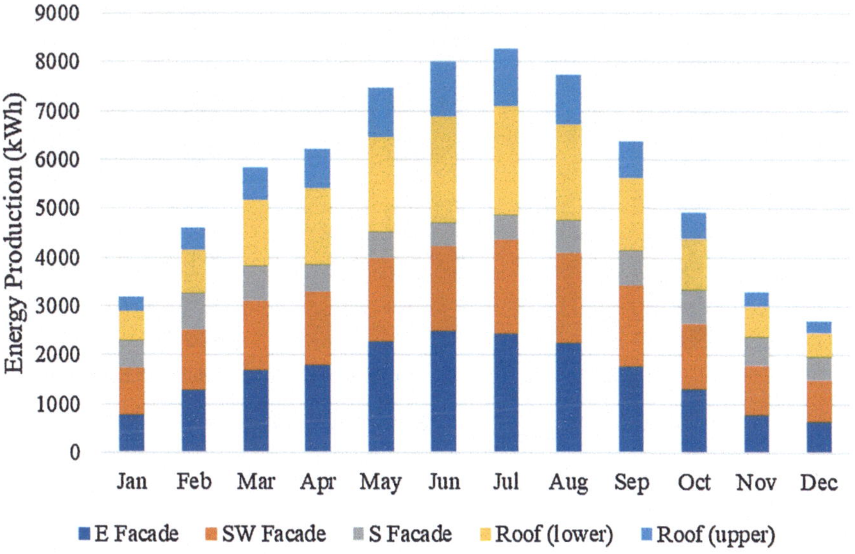

Fig. 11 Monthly electricity production by the BIPV

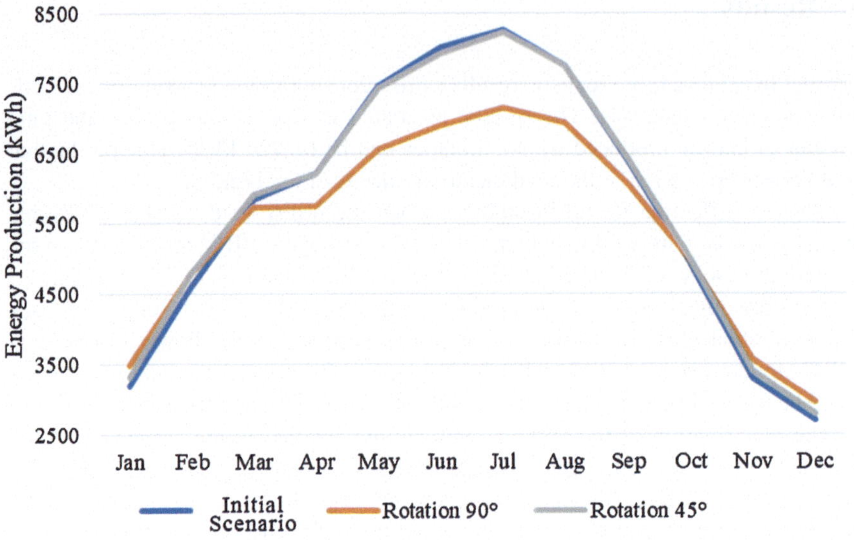

Fig. 12 Comparison between monthly electricity production by the BIPV for different orientations

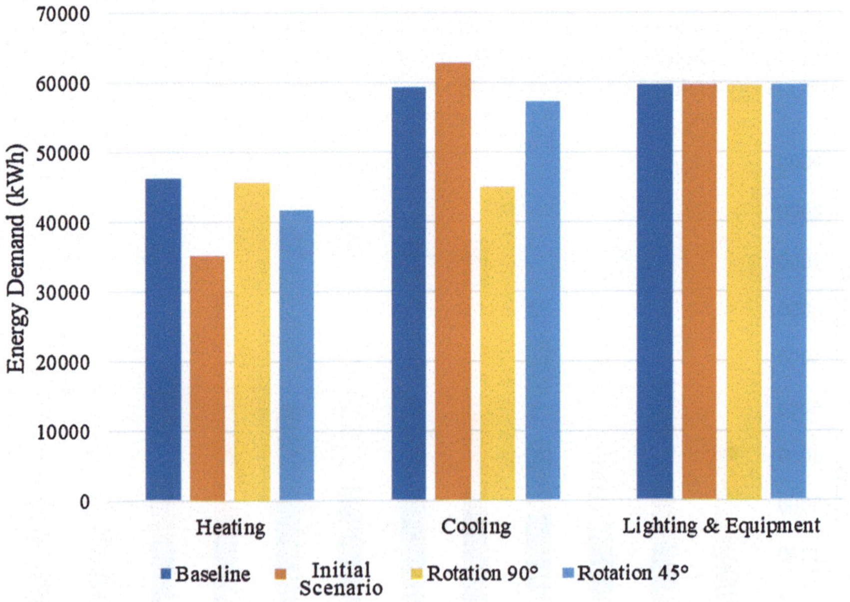

Fig. 13 Comparison between the annual energy demand of the scenarios

The energy demand of the building experiences important differentiations between the scenarios. The most energy intensive is, naturally, the baseline building, with a total energy demand of 159,074 kWh. As Fig. 13 displays, the rotation of 45° and the initial scenario are the ones that follow with a small deviation of about 1000 kWh between them (158,674 and 157,689 kWh respectively). The best scenario in terms of the energy demand is the 90° rotation, with approximately 15,000 kWh reduced demand compared to the baseline building.

7 Conclusions

The paper has focused on the retrofitting of one of the highest energy consumers of the building sector, office buildings, and their potential for energy refurbishment in order to achieve energy savings. The proposed retrofitting measures encompass higher performance building envelope and systems, with the integration of BIPV systems.

By assessing the energy performance of a typical nine-story office building in the center of Thessaloniki, the paper investigated the possibility of covering the energy demand of such buildings by solar energy. The evaluation of the building's existing condition and the potential of energy production and energy cost savings of the retrofitting scenarios proposed is achieved via energy simulations. The scenario provides promising results regarding the potential for electricity production by BIPV incorporated in facades and flat roofs, even in dense city centers, for buildings with a variety of typical orientations encountered throughout the cities. The amount of electricity produced however, is not enough to cover the whole energy demand of the office building, although the amortization period is attractive. The retrofitting should be more extensive in order for the energy consumption to be further reduced. Possible available measures could be an extended upgrading of the building's HVAC systems, and a modernization of the lighting fixtures and equipment. To that end the investigation of the cost optimal retrofitting scenario could be the scope of future research.

Additionally, the extended application of such interventions and the energy consumption reduction it will cause, can lead to a limitation of fossil fuel depletion and a decrease in air pollutants in urban areas. The cost of such interventions however, could render the implementation of retrofitting very challenging. The proposed retrofitting measures require a rather costly initial investment from the owners, and need an extended amortization period, showing however, a great potential.

Although there are significant barriers to be overcome in order to facilitate and accelerate BIPV technologies, especially regarding their cost-efficiency ratio, future work should continue to address the issue of office retrofitting with the incorporation of BIPV.

References

1. European Union. (2010). Directive 2010/31/EU of the European Parliament and of the Council of 19 May 2010 on the energy performance of buildings (recast). *Official Journal of the European Union, 153*, 13–35. https://doi.org/10.3000/17252555.L_2010.153.eng.
2. Martinopoulos, G., Papakostas, K. T., & Papadopoulos, A. M. (2018). A comparative review of heating systems in EU countries, based on efficiency and fuel cost. *Renewable and Sustainable Energy Reviews, 90*, 687–699.

3. European Commision. (2002). Directive 2002/91/EC of the European Parliament and of the Council of 16 December 2002 on the energy performance of buildings. *Official Journal of the European Union, 1*, 65–71. https://doi.org/10.1039/ap9842100196.
4. Artola, I., Rademaekers, K., Williams, R., & Yearwood, J. (2016). *Boosting building renovation: What potential and value for Europe*. Brussels: Policy Department A: Economic and Scientific Policy, European Parliament.
5. Economidou, M., Laustsen, J., Ruyssevelt, P., & Staniaszek, D. (2011). *Europe's buildings under the microscope*. Brussels: Buildings Performance Institute Europe (BPIE).
6. Sandels, C. (2016). *Modelling and simulation of electricity consumption profiles in the Northern European Building Stock* (Doctoral dissertation, KTH Royal Institute of Technology).
7. Zhang, T., Siebers, P. O., & Aickelin, U. (2011). Modelling electricity consumption in office buildings: An agent based approach. *Energy and Buildings, 43*(10), 2882–2892.
8. Martinopoulos, G., Papakostas, K. T., & Papadopoulos, A. M. (2016). Comparative analysis of various heating systems for residential buildings in Mediterranean climate. *Energy and Buildings, 124*, 79–87.
9. European Commission. (2016). *Impact assessment on Directive 2010/31/EU on the energy performance of buildings*. Retrieved from http://eur-lex.europa.eu/legalcontent/EN/TXT/PDF/?uri=CELEX:52016SC0414&from=EN.
10. BPIE. (2015). *Buildings modernisation strategy: Roadmap 2050* (p. 8). Retrieved from http://bpie.eu/wpcontent/uploads/2015/10/BuildingsModernisationStrategy2050ENsummary.pdf.
11. Atanasiu, B., Kouloumpi, I., Thomsen, K. E., Aggerholm, S., Enseling, A., Loga, T., & Witczak, K. (2013). *Implementing the cost-optimal methodology in EU countries: Lessons learned from three case studies*. Brussels: BPIE.
12. Martinopoulos, G. (2018). Life cycle assessment of solar energy conversion systems in energetic retrofitted buildings. *Journal of Building Engineering*. https://doi.org/10.1016/j.jobe.2018.07.027.
13. Johansson, P., & Wahlgren, P. (2017). Renovation of buildings from before 1945: Status assessment and energy efficiency measures. *Energy Procedia, 132*, 951–956.
14. Lassandro, P., & Di Turi, S. (2017). Façade retrofitting: From energy efficiency to climate change mitigation. *Energy Procedia, 140*, 182–193.
15. Kalogirou, S. A. (2001). Use a TRNSYS for modelling and simulation of a hybrid pv-thermal solar system for Cyprus. *Renewable Energy, 23*(2), 247–260.
16. Wigginton, M., & Harris, J. (2002). *Intelligent skins*. Oxford, UK: Architectural Press. Retrieved from http://medcontent.metapress.com/index/A65RM03P4874243N.pdf.
17. El Gindi, S., Abdin, A. R., & Hassan, A. (2017). Building integrated photovoltaic retrofitting in office buildings. *Energy Procedia, 115*, 239–252.
18. Tsalikis, G., & Martinopoulos, G. (2015). Solar energy systems potential for nearly net zero energy residential buildings. *Solar Energy, 115*, 743–756.
19. Gazeas, S., Gouzkounis, A., & Tzekakis, E. (2016). Architectural criteria for building integrated photovoltaics in building envelopes (in Greek). *ECON3*. Retrieved September 8, 2017, from http://www.econ3.gr/readmore.php?article_id=40311295559954.
20. Becchio, C., Corgnati, S. P., Vio, M., Crespi, G., Prendin, L., Ranieri, M., & Vidotto, D. (2017). Toward NZEB by optimizing HVAC system configuration in different climates. *Energy Procedia, 140*, 115–126.
21. Wemhoener, C., Schwarz, R., & Rominger, L. (2017). IEA HPT Annex 49-Design and integration of heat pumps in nZEB. *Energy Procedia, 122*, 661–666.
22. Hellenic Statistical Authority. (2015). *2011 General census of buildings* (in Greek). Athens, Greece: Hellenic Statistical Authority.
23. Hellenic Statistical Authority. (2007). *Population—Housing census results of 18th March 2001* (in Greek). Athens, Greece: Hellenic Statistical Authority.
24. Greek Parliament. (1979). *Thermal insulation regulation* (in Greek). Official Gazette of the Hellenic Republic, D' 362.
25. Greek Parliament. (2010). *Regulation for the energy performance of buildings (K.EN.A.K.)* (in Greek) (pp. 5333–5356). Official Gazette of the Hellenic Republic.

26. Ministry of Environment and Energy. (2012). *Statistical data of Energy Performance Certificates* (in Greek). Special Agency of Energy Efficiency Auditors.
27. Dascalaki, E. G., Droutsa, K. G., Balaras, C. A., & Kontoyiannidis, S. (2011). Building typologies as a tool for assessing the energy performance of residential buildings—A case study for the Hellenic building stock. *Energy and Buildings, 43*(12), 3400–3409.
28. Theodoridou, I., Papadopoulos, A. M., & Hegger, M. (2011). A typological classification of the Greek residential building stock. *Energy and Buildings, 43*(10), 2779–2787.
29. Greek Parliament. (1955). *National general building regulation* (in Greek). Official Gazette of the Hellenic Republic, A' 266.
30. European Commission. (2014). *Communication from the Commission to the European Parliament, the Council, the European Economic and Social Committee and the Committee of the Regions, Energy prices and costs in Europe*. COM (2014) 021 final.
31. European Parliament. (2012). Directive 2012/27/EU. Official Journal of the European Union, L315/1pp. 1–56. doi: https://doi.org/10.3000/19770677.L_2012.315.eng.
32. Becchio, C., Corgnati, S. P., Vio, M., Crespi, G., Prendin, L., & Magagnini, M. (2017). HVAC solutions for energy retrofitted hotel in Mediterranean area. *Energy Procedia, 133*, 145–157.
33. Cetiner, I., & Özkan, E. (2005). An approach for the evaluation of energy and cost efficiency of glass façades. *Energy and Buildings, 37*(6), 673–684.
34. EnergyPlus. (2017). *EnergyPlus*. Retrieved November 1, 2017, from https://energyplus.net/.
35. Strand, R., Winkelmann, F., Buhl, F., Huang, J., Liesen, R., Pedersen, C., Fisher, D., Taylor, R., Crawley, D., & Lawrie, L. (1999). Enhancing and extending the capabilities of the building heat balance simulation technique for use in EnergyPlus. *Building Simulation, 2*, 653–660.
36. Trimble. (2017). *SketchUp*. Retrieved December 2, 2017, from https://www.sketchup.com/.
37. Big Ladder Software. (2017). *Euclid*. Retrieved December 2, 2017, from https://bigladdersoftware.com/projects/euclid/.
38. National Renewable Energy Laboratory. (2005). *System Advisory Model (SAM), 2005*. Retrieved September 5, 2017, from https://sam.nrel.gov/.
39. Blair, N., Dobos, A. P., Freeman, J., Neises, T., & Wagner M. (2014). *System advisor model, sam 2014.1. 14: General description* (p. 13). NREL Report No. TP-6A20-61019. Golden, CO: National Renewable Energy Laboratory. Retrieved from http://www.nrel.gov/docs/fy14osti/61019.pdf.
40. Cameron, C. P., Boyson, W. E., & Riley, D. M. (2008). Comparison of PV system performance-model predictions with measured PV system performance. In *Conference Record of the IEEE Photovoltaic Specialists Conference*. https://doi.org/10.1109/PVSC.2008.4922865.
41. De Andalucia, F. (2013). *System Advisor Model (SAM) case study: Gemasolar* (pp. 1–10). Nrel. Retrieved from https://sam.nrel.gov/sites/sam.nrel.gov/files/content/case_studies/sam_case_csp_physical_trough_andasol-1_2013-1-15.pdf.
42. Serasidou, A., Petousis, E., & Martinopoulos, G. (2018). Building Integrated Renewable Energy Systems for energy retrofitting of a single-family residence towards an nZEB (in Greek). In *11th National Conference on Renewable Energy Sources*, Thessaloniki.
43. American Society of Heating Refrigerating and Air-conditioning Engineers (ASHRAE). (2013). ASHRAE Standard 55-2013 Thermal environmental conditions for human occupancy. *Ashrae, 2004*.
44. Gan, G. (2009). Numerical determination of adequate air gaps for building-integrated photovoltaics. *Solar Energy, 83*(8), 1253–1273.
45. Ministry of Environment and Energy. (2014). *Installation of RES under the net metering scheme according to article14A of the law 3468/2006* (in Greek). Official Gazette of the Hellenic Republic, A' 129.
46. Staff I. (2018). *Payback period*. Investopedia. Retrieved September 10, 2017, from https://www.investopedia.com/terms/p/paybackperiod.asp.
47. Public Power Corporation. (2017). *P.P.C.—Pricing for companies—C21*. Retrieved September 10, 2017, from https://www.dei.gr.

Electric Lighting Predictions in the Energy Calculation Methods

Francesco De Luca, Raimo Simson, Hendrik Voll, and Jarek Kurnitski

Abstract Electric lighting is one of the major factors for energy consumption of buildings. The European Directive 2010/31/EU states that from 31 December 2020 all new buildings will have to be nearly-Zero Energy Buildings, thus improving electric lighting energy performance is a key issue. The article presents a study and energy figures of power density and electric lighting annual consumptions for different types of buildings, office, commercial and educational, in the northern European country Estonia with the scope to quantify energy savings when using different types of high-efficiency luminaires, occupancy and dimming controls, lighting groups, and daylight contribution. The study has been conducted in relation to the energy performance regulation for new buildings in Estonia. The scope is to develop methods for electric lighting and daylight calculations to be used in compliance assessment with energy requirements. Using different validated software for electric light and daylight simulations the study analyzes three cases for office buildings, single office, open office and meeting room, and one case for both commercial and educational buildings. Results show that average installed power density can be as low as 3.17 W/m^2 for office rooms, 3.22 W/m^2 for commercial buildings and 2.09 W/m^2 for classrooms. The reduction of energy consumption comparing tabulated values can be up 93.3% for office rooms. Also for commercial and educational buildings energy saving are consistent, up to 72.2% and 87.2% respectively. The article presents as well electric light and daylight model specifications and parameters and the different control settings and relative performance.

F. De Luca (✉) · R. Simson · H. Voll
Department of Civil Engineering and Architecture, Tallinn University of Technology, Tallinn, Estonia
e-mail: francesco.deluca@taltech.ee

J. Kurnitski
Department of Civil Engineering and Architecture, Tallinn University of Technology, Tallinn, Estonia

Department of Civil Engineering, Aalto University, Espoo, Finland

P. Bertoldi (ed.), *Improving Energy Efficiency in Commercial Buildings and Smart Communities*, Springer Proceedings in Energy,
https://doi.org/10.1007/978-3-030-31459-0_9

1 Introduction

Nearly-Zero Energy Buildings performance depend on a large quantity of factors that need to be carefully studied and analyzed during the design phase. All European Union countries are required to set the target to improve energy efficiency in new constructions on the basis of the European Directive on energy performance of buildings 2010/31/EU that states that by from 31 December 2020 all the new buildings will have to be nearly-Zero Energy Buildings [1].

In the Northern European country Estonia the energy consumption is regulated by the ordinance "Minimum Requirements for Energy Performance" that states that nearly-Zero Energy Buildings of the type multi-apartment and office building cannot consume more than 100 kWh/m^{2}y, business and commerce buildings not more than 130 kWh/m^{2}y and educational buildings not more than 90 kWh/m^{2}y of primary energy [2].

Electric lighting is responsible for energy use for up to 40% of total consumption in commercial and public buildings, thus relying huge potentialities of energy saving [3–6]. The total electricity used in commercial buildings on a global scale is used mostly by office, commercial, and educational buildings with 19%, 25%, and 10% respectively [7]. This make saving energy for electric lighting a key strategy for the total energy efficiency of commercial buildings and an effective way to fulfill the European Directive 2010/30/EU.

Are several the strategies that can significantly decrease the electric lighting energy consumption, that is, the utilization of high-efficiency lamps and improved performance luminaires, division of the space in task and surrounding areas, occupancy sensors and controls, daylight dimming, and lighting groups.

High-efficiency lamps are the first strategy to reduce lighting energy consumption. T5 fluorescent lamps with an high efficacy of about 100 lm/W permit to have a reduction of energy use up to 40% comparing older lamp technology with an efficacy as low as 60 lm/W [8]. LED lamps with higher efficacy of 160 lm/W are nowadays available, with room for further improvements, at costs that are competitive with conventional lighting considering the installation costs but that are significantly smaller considering the entire luminaires life cycle [9].

The distinction of the interior spaces of commercial buildings into areas with different target maintained illuminance, that is, task areas and secondary building areas and amenities, is a valid strategy to reduce electric light energy consumption and at the same time guarantee adequate illumination for the specific tasks [10]. The European Lighting Standard EN 12464-1:2011 recommends the design of three different areas inside the same office room, retail space or class room, the task area relative to the main activity, that is, the desk surface of an office room, the immediate surrounding area with a depth of at least 0.5 m, and a background area of at least 3 m or up to room walls, with different maintained illuminance required [11].

Many studies have been conducted about potential electric light energy saving due to manual dimming and automatic on/off controls driven by occupancy sensors. A research carried on during an entire year in an open plan office using different controls, on/off occupancy sensors, light sensors and individual dimming controls showed that energy savings can be in the range of 42–47% if compared to the same high-efficacy lamps used and in the range of 67–69% if compared to lower efficacy lamps [12].

Daylight is one of the main factors for electric light energy saving in commercial, public, and retail spaces [13]. In today's nearly-Zero Energy Buildings high-efficiency luminaires controlled by daylight controls and illuminance sensors are a compulsory strategy [14–16].

The use of sensors and controllers associated with daylight availability can dramatically reduce or even eliminate the need of electric lighting depending on the location of the building and period of the year [17]. The use of high-efficiency LED lamps can increase the savings that is possible to obtain using daylighting and automatic dimming controls due to the capacity of LED luminaires that include efficient drivers to be dimmed down to very low power values [18–20].

Daylighting harvesting to reduce electric light energy consumption has to be carefully studied since it can increase the need of cooling and heating energy on an annual basis [21, 22]. Nevertheless, exterior static or internal manual or dynamic shading systems if properly designed can minimize higher cooling and heating loads and possible glare effects [23, 24].

The use of daylighting as an effective strategy to reduce electric light energy consumption can be maximized with a careful division of the daylit area in different lighting zones with different groups of luminaires with separate sensors and controls. This strategy proved to save an additional range from 14% to 39% of electric lighting in commercial buildings with daylight comparing the utilization of single lighting zone [25].

The present study investigates the minimum power density and relative annual consumptions using luminaires with different efficiency, occupancy controls, daylight dimming, and lighting groups to guarantee the maintained illuminance as required by the European Lighting Standard EN 12464-1:2011 for the building type office, commercial, and educational in the city of Tallinn, Estonia. The research has been conducted in relation to the new Electric Light Guidance developed to provide electric light designers a procedure about how to select the most energy efficient lighting plan in their surveys as part of the energy calculations for the Estonian Building Code, Building Energy Efficiency charter, and how to assess the efficiency of the selected plan in relation to tabulated values of the typical use of buildings [26].

2 Methodology

The methods related to the study are two. The first is the method developed as a set of procedures to be used by electric light designers in Estonia for the assessment of the energy efficiency of their electric lighting plans. This is also the first outcome of the research and its main scope. The second is the method used to investigate the potentialities of electric light energy reduction when using energy saving strategies in the city of Tallinn, Estonia, for different types of commercial and public buildings. This is also part of the developed guidance as examples of the surveys required to the electric light designers.

2.1 Electric Lighting Calculations

The Electric Light Guidance to be used in relation to the Estonian Building Code, Building Energy Efficiency charter, is based on the recommendations of the European Lighting Standard EN 12464-1:2011, that is, the maintained illuminance and uniformity values for the different type of buildings and the division of the space in task and secondary areas. Additionally the guidance expands its content suggesting energy efficiency strategies, that is, the use of occupancy sensors, daylight dimming, and lighting zones.

The lighting requirements indicated are relative to specific type of rooms (i.e., circulation and corridors, meeting rooms, classrooms) and building types (i.e. general areas, offices, and educational). As indicated by the standard for each type of room is set the required average maintained illuminance (lx) and uniformity (U_0).

The Electric Light Guidance recommends procedures for the assessment through computer simulations of electric light consumption, with the scope to select efficient lamps and luminaires by the electric light designers, for the building types included in the European Lighting Standard such as office and commercial. For other type of buildings such as residential, not included in the standard to determine the annual electric light energy consumption, the typical use of buildings tabulated data are used [26].

For the calculation of the annual energy consumption the presented formula (1) is used. Additionally room usage rate is provided in the guidance.

$$Q = kP\frac{\tau_{\mathrm{d}}}{24}\frac{\tau_{\mathrm{w}}}{7}\frac{8760}{1000} \tag{1}$$

Q energy consumption (kWh/m^{2}y)
K room usage rate
P installed power density
τ_{d} number of operating hours per day (h)
τ_{w} number of days of use per week (day)

For the majority of the building types such as commercial, office, and educational it is required the determination of installed power and annual energy consumption through computer simulation. The first step is to select the most relevant rooms of the analyzed building as indicated in the guidance. Once determined the annual energy consumption for that type of room the total energy consumption of the part of the building occupied by those rooms is calculated.

For the determination of the installed power and relative power density a simulation software is used, that is, DIALux, Relux, and Radiance, that can accurately calculate the illuminance levels and uniformity values using lamps and luminaires photometric data from digital and online catalogues.

The guidance recommends the utilization of high-efficiency and low-consumption lamps and luminaires such as LED as a first strategy to reduce electric light energy consumption.

The calculation areas are those indicated by the European Lighting Standard EN 12464-1:2011: the task area, the surrounding area and the background area, located at a height of 0.8 m from the floor.

The guidance follows the European standard for the reflectance coefficient of the interior finishing to be used in the simulations and for the recommended lamps maintenance factors.

2.2 Determination of Annual Energy Consumption

The scope of the Electric Light Guidance is that to provide the electric light designer with the procedures to correctly assess the electric light energy consumption as part of the whole energy assessment of buildings to certify the nearly-Zero Energy Building requirement. At the same time the guidance recommends the following strategies for the energy efficiency of electric lighting:

- Daylight contribution
- Occupancy controls
- Lighting controls
- Lighting groups

As discussed in Sect. 1 daylight can significantly reduce the need of electric lighting. The developed guidance presents methods and settings for the calculation of annual electric light energy consumption that consider daylight contribution and controls used by the simulation software IDA-ICE and Radiance/Daysim. Other software can be used with similar methods and settings.

The daylight simulation software calculates the percentage of time, during the occupied hours for the entire year, during which daylight is enough or electric light is necessary, and in case of dimmable luminaires the ratio of necessary electric light, to reach the required illuminance levels. Consequently the software multiply the time (hours) for the installed power or a ratio of it. The result is the annual electric light energy consumption. The metrics used by the software Radiance/Daysim in the two cases, without or with dimmable luminaires, are the Daylight Autonomy (DA) and Continuous Daylight Autonomy (CDA) [27]. The simulation software predicts the daylight contribution on the basis of statistical weather data of the specific latitude and longitude of the location.

During building operation occupancy sensors can significantly reduce the use of electric lighting switching off light when the space is not occupied, that is, during off hours in case occupants leave light on. During the design assessment phase daylight simulation software calculates the percentage of time during which daylight is enough to guarantee the required illuminance levels and the quantity of necessary electric light to guarantee the required illuminance during the hours when the room is occupied every day of the year it is necessary to light the room (i.e., for office weekdays 9–17).

Besides the consideration of the time during which the room is occupied or not, it is possible to use different types of controls that mimic the behavior of the occupants in controlling the electric light when there is availability of daylight [28]. The main types of controls for electric light recommended by the guide are presented in Table 1.

For a higher accuracy of the electric light energy annual consumption and above all in the case of large rooms in which the difference of daylight is significant the guidance recommends the use of different lighting groups for different zones of the room. The reason is that if only a sensor node is used in a large room, if the sensor node is located in a well daylit zone it will determine that also not well daylit zones in the room will have the light dimmed or switched off when it would be necessary, or if the sensor node is located in a not well daylit zone it will determine that also well daylit zones will have the light switched on when it would not be necessary.

The Electric Light Guidance shows the procedure to perform surveys of solutions that integrate controls and daylighting and their comparison by the electric light designer to select the most efficient electric light plan and strategy as required by the Estonian Building Code, Building Energy Efficiency charter.

For each selected room type three electric lighting design solutions are studied and compared with the tabulated values of the typical use of buildings. The three

Table 1 Lighting control types

Control		Comment	Simulation software
1	Set-points and schedule	Control that uses the lighting set-point (lx) and the occupancy schedule selected.	IDA-ICE
2	Manual On/Off Switch	Control that mimics the behavior of occupants in switching off the light when there is enough daylight illuminance on the desktop.	Radiance/ Daysim
3	Switch Off with Occupancy	Control that mimics the behavior of occupants in switching off the light when there is enough daylight illuminance on the desktop and makes sure that occupants do not leave the light on when they are not in the room during off hours.	
4	Photosensor Controlled Dimming	Control that uses a continuous dimming sensor in relation to the available daylight.	
5	Dimming with Occupancy Off Sensor	Control that uses a continuous dimming sensor in relation to the available daylight and that makes sure that occupants do not leave the light on when they are not in the room during off hours.	

Table 2 Average maintained illuminance and uniformity recommended by the European Standard EN12464-1

Building type	Area type	Average maintained illuminance (lx)	Uniformity (U_0)
Office	Task area	500	0.6
	Surrounding area	300	0.4
	Background area	100	0.1
Educational	Task area	300	0.6
	Surrounding area	200	0.4
	Background area	66.7	0.1
Commercial	Task area	500/300 (sales area/till area)	0.6
	Surrounding area	300/200 (sales area/till area)	0.4
	Background area	100/66.7 (sales area/till area)	0.1

Table 3 Data of material properties for electric light and daylight simulations

Property	Walls	Ceiling	Floor
Surfaces reflection coefficient (%)	50	70	20
Window glass visible transmittance (%)	70		

solutions are different for the type of luminaire selected, thus for the power density, and control used but all need to fulfill the requirements of illuminance levels and uniformity. For all the solutions the annual electric energy consumption analysis that includes daylight simulation and controls is done. The solution with the smallest energy consumption is therefore selected.

The recommended maintained illuminance and uniformity for each area of the different building types, used in the simulations for the electric light plans are presented in Table 2.

For all the electric lighting design solutions the simulation data of material properties used are presented in Table 3.

3 Results

The present section describes the room parameters, the luminaire and control options and the results for the five selected room type to investigate the potentialities of electric light energy saving strategies. The results of the examples are specific for the location of Tallinn, Estonia and for the used room sizes, interior finishing, window sizes, orientation, and material properties. Different rooms and characteristics would generate different results.

3.1 Single Office

The single office room has a size of 3 m × 4.5 m, a height of 2.8 m, and a window size of 2 m × 1.5 m oriented toward south (Fig. 1). The resultant Window to Wall ratio is 37%. The task area, the surrounding area and the background area are located at the desk height of 0.80 m from the floor and follow to the recommendations of the European Lighting Standard EN12464-1. The single office is located in Tallinn; therefore, the climate file for the city of Tallinn has been used.

The three options for the luminaires and controls and the results of the power density, task area average illuminance, task area uniformity, and the annual electricity consumption including daylighting for the three design options are presented in Table 4. The single office has only one lighting group.

The option with the least energy consumption (Option 3) is selected. The luminaire are two ceiling mounted high-efficiency dimmable LED (47.5 W) with a maintenance factor of 0.8, a driver loss factor of 0%, and a maximum dimming of 1%. The occupancy schedule used is Weekdays 09–17.

3.2 Open Office

The open office room has a size of 12 m × 6 m, a height of 2.8 m and four windows with size of 2 m × 1.35 m oriented toward south (Fig. 2). The resultant Window to Wall ratio is 32%. Has been used a task area and surrounding area for every working

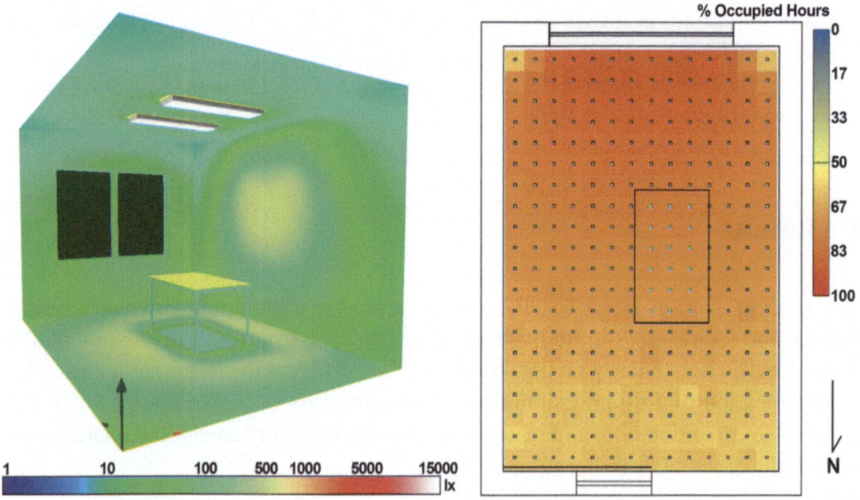

Fig. 1 Left—Electric lighting simulations using the software DIALux Evo (Values in lx). Right—Daylight simulation using software Radiance/Daysim (Mean Continuous Daylight Autonomy 500 lx 73% of occupied time)

Table 4 Results of electric light calculations and annual energy consumption with daylight

Options	Lamp type	Power density (W/m^2)	Average illuminance (lx)	Uniformity (U_0)	Control type	Annual electric consumption (kWh/m^2 year)
Tabulated value	–	12.0	–	–	–	18.91
1	TL5 38W × 4	11.2	531	0.96	Manual on/off switch	7.36
2	LED 41W × 3	9.1	523	0.96	Switch Off with Occupancy	4.10
3	LED 47.5W × 2	7.04	512	0.94	Dimming with Occupancy Off Sensor	2.49

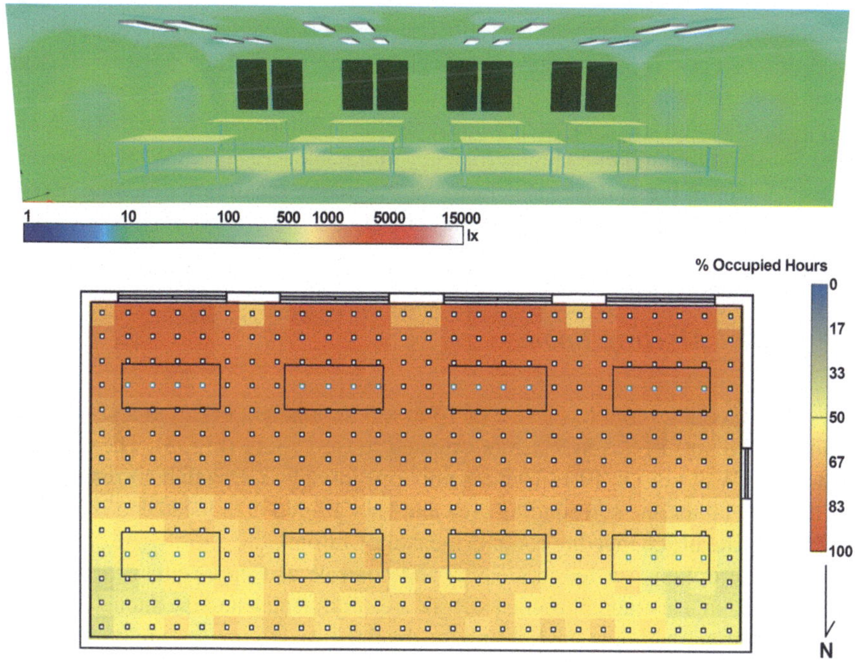

Fig. 2 Above—Electric lighting simulations (software DIALux Evo—Values in lx). Below—Daylight simulation (software Radiance/Daysim—Mean CDA 500 lx 68% of occupied time)

desk, at the height of 0.80 m, and the background area is common for all the desks. The open office is located in Tallinn; therefore, the climate file for the city of Tallinn has been used.

The three options for the luminaires and controls and the results of the power density, smallest maintained average illuminance among all task areas, relative task

Table 5 Results of electric light calculations and energy consumption with daylight

Options	Lamp type	Power density (W/m^2)	Average (min) illuminance (lx)	Uniformity (U_0)	Control type	Lighting groups	Energy use (kWh/m^2 year)
Tabulated value	–	12.0	–	–	–	–	18.91
1	TL5 100W × 8	11.1	509	0.83	Manual On/Off Switch	1–2	10.25–8.23
2	LED 41W × 16	9.11	519	0.86	Switch Off with Occupancy	1–2	6.26–5.04
3	LED 24.5W × 16	5.44	505	0.86	Dimming with Occupancy Off Sensor	1–2	3.06–2.37

area uniformity and the annual electricity consumption including daylighting for the three design options are presented in Table 5. The open office has two lighting groups corresponding to the two rows of desks parallel to the windows.

The option with the least energy consumption (Option 3) is selected. The luminaire are 16 ceiling mounted high-efficiency dimmable LED (24.5 W) with a maintenance factor of 0.8, a driver loss factor of 0%, and a maximum dimming of 1%. The occupancy schedule used is Weekdays 09–17.

3.3 *Office Meeting Room*

The office meeting room has a size of 10 m × 6 m, a height of 2.8 m and one long window with size of 7.5 m × 1.5 m oriented toward south (Fig. 3). The resultant Window to Wall ratio is 40%. There is only one large task area that corresponds to the meeting table, one surrounding area and background area located at the height of 0.80 m. The office meeting room is located in Tallinn.

The three options for the luminaires and controls and the results of the power density, task area average illuminance, task area uniformity, and the annual electricity consumption including daylighting for the three design options are presented in Table 6. The office meeting room has only one lighting group for all the three options.

The option with the least energy consumption (Option 3) is selected. The luminaire are four ceiling mounted high-efficiency dimmable LED (47.5 W) with a maintenance factor of 0.8, a driver loss factor of 0%, and a maximum dimming of 1%. The occupancy schedule used is Weekdays 09–17.

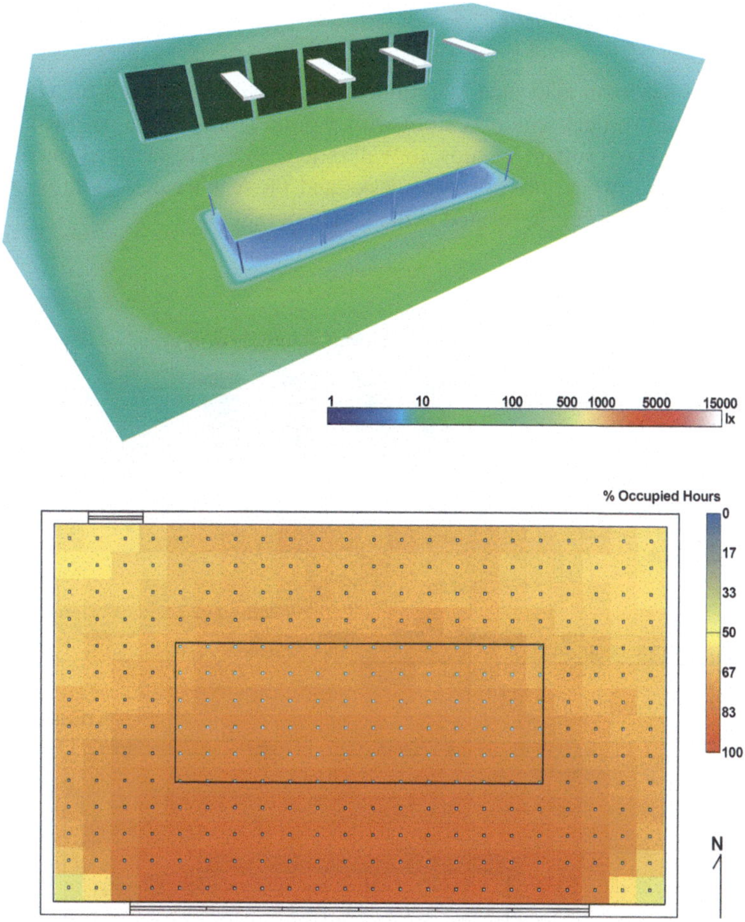

Fig. 3 Above—Electric lighting simulations (software DIALux Evo—Values in lx). Below—Daylight simulation (software Radiance/Daysim—Mean CDA 500 lx 73% of occupied time)

Table 6 Results of electric light calculations and energy consumption with daylight

Options	Lamp type	Power density (W/m^2)	Average illuminance (lx)	Uniformity (U_0)	Control type	Energy use (kWh/ m^2 year)
Tabulated value	–	12.0	–	–	–	18.91
1	TL5 100W × 3	5	509	0.84	Manual On/Off Switch	3.60
2	LED 41W × 6	4.1	509	0.74	Switch Off with Occupancy	2.06
3	LED 47.5W × 4	3.17	519	0.6	Dimming with Occupancy Off Sensor	1.26

3.4 Educational Room

The educational room is located in Tallinn and has a size of 9 m × 8 m, a height of 2.8 m and four vertical windows from with size of 1 m × 2.4 m oriented toward south (Fig. 4). The resultant Window to Wall ratio is 38%. The task and surrounding areas correspond to the three rows of desks and the teacher desk with one background area in common, all located at the height of 0.80 m.

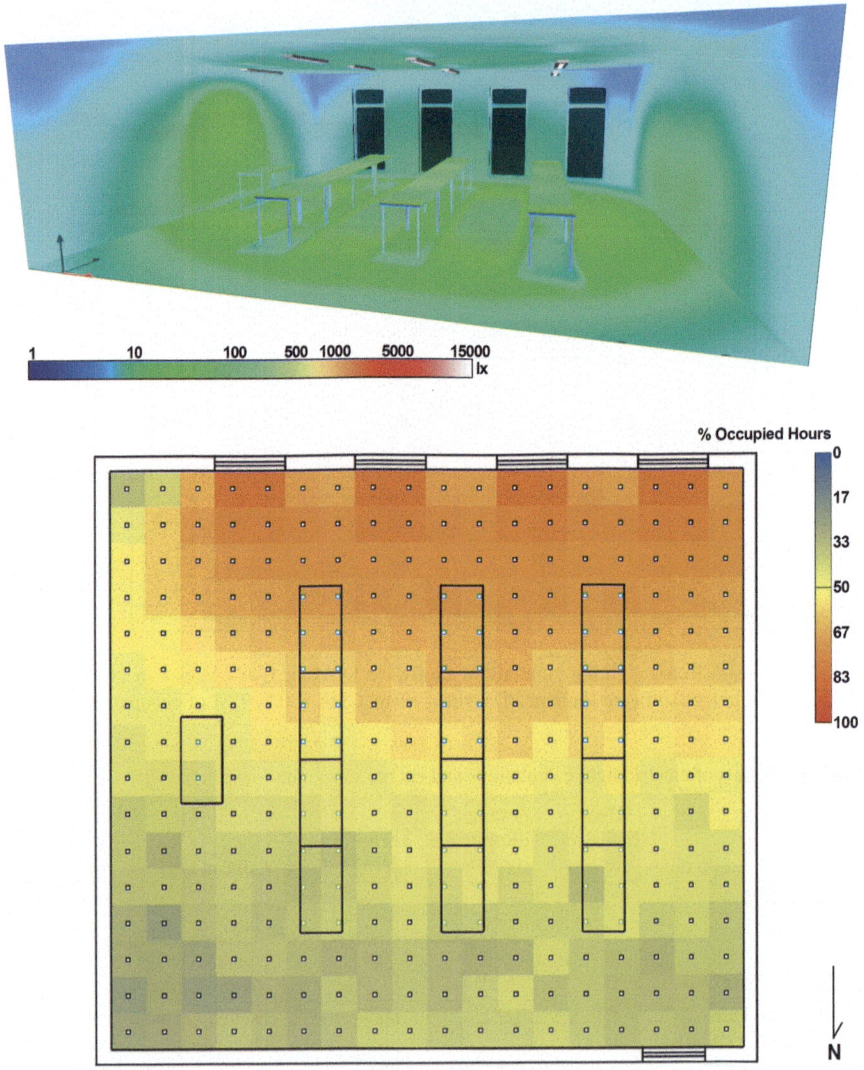

Fig. 4 Above—Electric lighting simulations (software DIALux Evo—Values in lx). Below—Daylight simulation (software Radiance/Daysim—Mean CDA 500 lx 52% of occupied time)

One lighting group for option 1 and 2 (teacher and students cannot switch light on and off regularly depending on daylight), and one and three lighting groups for option 3 have been used. The three lighting groups are the desks closer the windows, the teacher desk and the desks far from the windows. Results of the smallest average illuminance among the task areas, uniformities and consumptions are presented in Table 7.

The option with the least energy consumption (Option 3) is selected. The luminaire are seven suspended (20–50 cm) high-efficiency dimmable LED (21.5 W) with a maintenance factor of 0.8, a driver loss factor of 0%, and a maximum dimming of 1%. The occupancy schedule used is Weekdays 08–16.

3.5 *Commercial Room*

The commercial room is composed of two adjacent areas with size 30 m × 15 m for the sales and 15 m × 5 m for the tills area, both with a height of 4 m (Fig. 5). The facade of 15 m of the till area is totally glazed with windows of height 2.5 m oriented toward south. The example commercial room is located in Tallinn; therefore, the climate file for the city of Tallinn has been used.

The three options for the luminaires and controls and the results of the power density, task areas average illuminance, task areas uniformity, and the annual electricity consumption including daylighting for the three design options are presented in Table 8. For the commercial room have been used two lighting groups for all the three options. No occupancy off control has been used since in a commercial activity the light is switched off at night by security personnel.

The option with the least energy consumption (Option 3) is selected. The luminaire are 69 suspended high-efficiency dimmable LED (24.5 W) with a maintenance

Table 7 Results of electric light calculations and energy consumption with daylight

Options	Lamp type	Power density (W/m^2)	Average (min) illuminance (lx)	Uniformity (U_0)	Control type	Lighting groups	Energy use (kWh/m^2 year)
Tabulated value	–	15.0	–	–	–	–	15.64
1	TL5 29W × 10	4.03	318	0.86	Manual On/Off Switch	1	6.17
2	LED 24.5W × 7	2.38	319	0.66	Switch Off with Occupancy	1	3.71
3	LED 21.5W × 7	2.09	300	0.62	Dimming with Occupancy Off Sensor	1–3	2.22–1.99

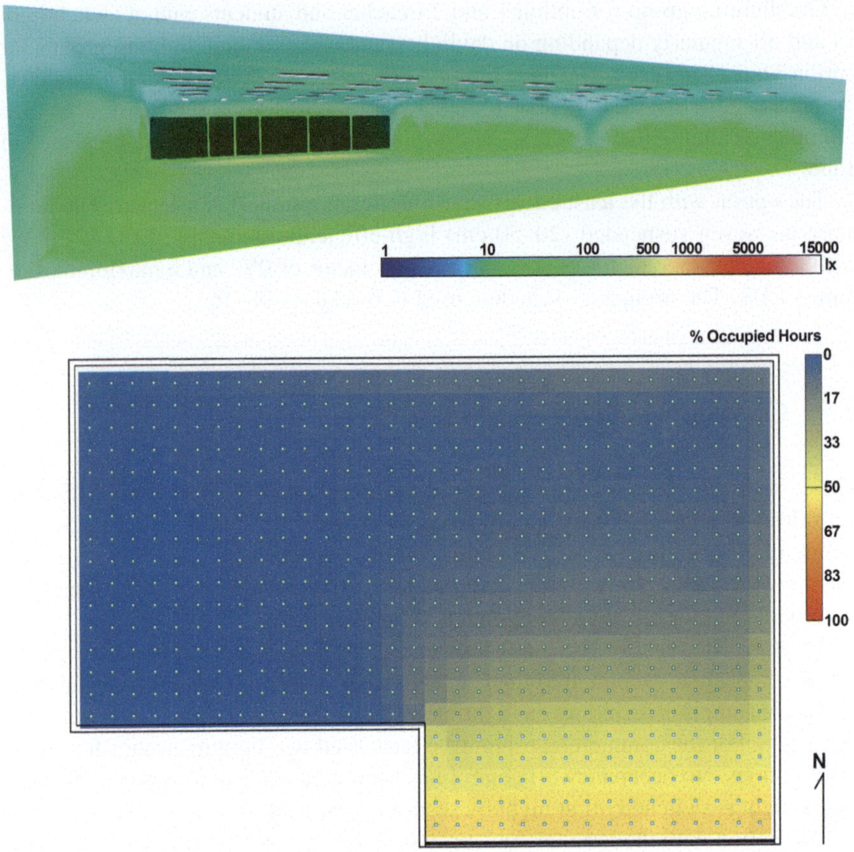

Fig. 5 Above—Electric lighting simulations (software DIALux Evo—Values in lx). Below—Daylight simulation (software Radiance/Daysim—Mean CDA 500 lx 13% of occupied time)

Table 8 Results of electric light calculations and energy consumption with daylight

Options	Lamp type	Power density (W/m^2)	Average illuminance (lx) Sales-tills area	Uniformity (U_0) Sales-tills area	Control type	Energy use (kWh/m^2 year)
Tabulated value	–	20.0	–	–	–	56.18
1	TL5 110W × 32	6.70	569–334	0.73–0.66	Manual On/Off Switch	31.7
2	LED 41W × 68	5.31	543–349	0.73/0.52	Manual On/Off Switch	25.6
3	LED 24.5W × 69	3.22	501–330	0.71/0.66	Photosensor Controlled Dimming	15.62

factor of 0.8, a driver loss factor of 0%, and a maximum dimming of 1%. The occupancy schedule used is Whole week 09–23.

3.6 Summary of Results

A summary of results is presented in Fig. 6. For the single office room using high-efficiency lamps and luminaires the power densities are 11.20, 9.10, and 7.04 W/m^2 that corresponds to 93.3%, 75.8%, and 58.6% of the power density of 12 W/m^2 of the tabulated values for office rooms when using fluorescent TL5 lamps and two types of LED luminaires with increasing efficacy respectively. If lighting controls and daylight dimming are accounted for, the consumption is 38.9%, 21.6%, and 13.1% using the three mentioned luminaires and Manual On/Off Switch, Switch Off with Occupancy, and Dimming with Occupancy Off Sensor controls respectively comparing the reference consumption of 18.91 kWh/m^{2}y obtained with power density of the tabulated values and the formula presented in Sect. 2.1. All the simulations for the single office room have been conducted with one lighting group due to the presence in the room of only one task area.

For the open office room using high-efficiency lamps and luminaires the power densities are 11.10, 9.11, and 5.44 W/m^2 that corresponds to 92.5%, 75.9%, and 45.3% of the power density of 12 W/m^2 of the tabulated values for office rooms when using fluorescent TL5 lamps and two types of LED luminaires with increasing efficacy respectively. In the open office room simulations have been used one and two lighting groups. If lighting controls, daylight dimming, and lighting groups are accounted for the electric light consumption using the three mentioned luminaires and Manual On/Off Switch, Switch Off with Occupancy, and Dimming with

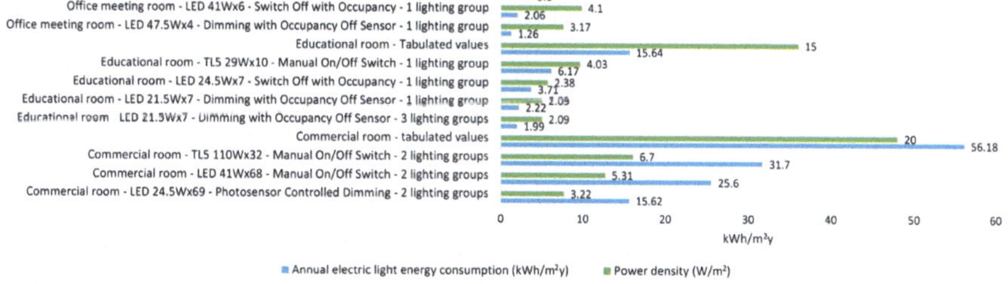

Fig. 6 Summary of results for power densities and annual electric light energy consumptions for the different rooms and options and comparison with tabulated values

Occupancy Off Sensor controls respectively is 54.2%, 33.1%, and 16.1% of the reference consumption of 18.91 kWh/m^{2}y. In the case of two lighting groups the consumption is 54.2%, 26.6%, and 12.5% respectively of the reference consumption.

For the office meeting room using high-efficiency lamps and luminaires the power densities are 5.0, 4.10, and 3.17 W/m^2 that corresponds to 41.6%, 34.1%, and 26.4% of the power density of 12 W/m^2 of the tabulated values for office rooms when using fluorescent TL5 lamps and two types of LED luminaires with increasing efficacy respectively. If lighting controls and daylight dimming are accounted for, the electric light energy consumption using the three previously mentioned luminaires and Manual On/Off Switch, Switch Off with Occupancy, and Dimming with Occupancy Off Sensor controls respectively is 19.0%, 10.9%, and 6.66% of the reference consumption of 18.91 kWh/m^{2}y obtained with power density of the tabulated values.

For the educational room using high-efficiency lamps and luminaires the power densities are 4.03, 2.38, and 2.09 W/m^2 that corresponds to 26.8%, 15.8% and 13.9% of the power density of 15 W/m^2 of the tabulated values for educational rooms when using fluorescent TL5 lamps and two types of LED luminaires with increasing efficacy respectively. In the educational room simulations have been used one and three lighting groups as discussed in Sect. 3.4. In the case of one single lighting group using the three mentioned luminaires and Manual On/Off Switch, Switch Off with Occupancy, and Dimming with Occupancy Off Sensor controls respectively and in the case of three lighting groups for the third type of luminaire and control the consumption is 39.4%, 23.7%, 14.2%, and 12.7% for the four options comparing the reference consumption of 15.64 kWh/m^{2}y obtained with power density of the tabulated values.

For the commercial room using high-efficiency lamps and luminaires the power densities are 6.70, 5.31, and 3.22 W/m^2 that corresponds to 33.5%, 26.5%, and 16.1% of the power density of 20 W/m^2 of the tabulated values for commercial rooms when using fluorescent TL5 lamps and two types of LED luminaires with increasing efficacy respectively. If lighting controls, daylight dimming, and lighting groups are accounted for, the electric light energy consumption using the three previously mentioned luminaires and Manual On/Off Switch control for the first two options and Photosensor Controlled Dimming for the third option is 56.4%, 45.5%, and 27.8% of the reference consumption of 56.18 kWh/m^{2}y obtained with power density of the tabulated values and the formula presented in Sect. 2.1.

4 Conclusions

The present study confirms the previous research mentioned in Sect. 1 about the potentialities of energy saving strategies such as efficient lamps and luminaires, different illuminance requirement areas in the same room, occupancy controls, daylight dimming, and lighting groups.

Results show that installed power density can be as low as 7.04, 5.44, and 3.17 W/m^2 for office rooms, respectively single office, open plan office and meeting room, 2.09 W/m^2 for educational room and 3.22 W/m^2 for commercial room. The energy reduction considering all the mentioned electric light energy saving strategies comparing the tabulated values can be up to 86.8%, 87.4%, 93.3% respectively for the office room, single office, open plan office and meeting room, 87.2% for the educational room and 72.2% for the commercial room.

From the presented figures it is possible to conclude that the energy reduction of electric lighting can contribute significantly to reach the target of low energy buildings and for the total energy saving. To reach the nearly-Zero Energy Building level as set by Estonia regulations, real-estate developers, architects and engineers have to design and invest in different energy strategies such as high-efficiency lamps and luminaires, adequate electric light plan design that take into account the different task areas in the rooms, occupancy sensors, and daylight dimming sensors.

Future work of the discussed study is that to include whole building energy simulations to evaluate the incidence of solar gains when the use of daylight is maximized and the possible necessity to consider shading devices, and of internal gains that are lower with high-efficiency LED luminaires that is positive during the warm season because it permits to reduce cooling loads but negative during the cold months because it require an increase in heating energy.

Acknowledgments The research has been supported by the Estonian Centre of Excellence in Zero Energy and Resource Efficient Smart Buildings and Districts, ZEBE, grant 2014-2020.4.01.15-0016 funded by the European Regional Development Fund, under Institutional research funding grant IUT1-15 and by the Estonian Research Council with Personal research funding grant PUT-652.

Figure source and notes: All the figures have been realized by the authors.

References

1. European Commission. (2010). Commission Directive 2010/31/EU of the European Parliament and of the Council of 19 May 2010 on the energy performance of buildings. *Official Journal of the European Union, L 153, 53*, 13–35. Retrieved from https://eur-lex.europa.eu/.
2. Estonian Government. (2012). *Minimum requirements for energy performance of buildings* (Energiatõhususe miinimumnõuded). Passed 30.8.2012, Annex 68, Tallinn. Retrieved form https://www.riigiteataja.ee/en/eli/520102014001.
3. Krarti, M. (2010). *Energy audit of building systems: An engineering approach*. Boca Raton: CRC Press. ISBN 9781439828717.
4. Nazaroff, W. W. (2014). Illumination, lighting technologies, and indoor environmental quality. *Indoor Air, 24*(3), 225–226.
5. Ponmalar, V., & Ramesh, B. (2014). Energy efficient building design and estimation of energy savings from daylighting in Chennai. *Energy Engineering, 111*(4), 59–80.
6. Gago, E. J., Muneer, T., Knez, M., & Koster, H. (2015). Natural light controls and guides in buildings. Energy saving for electrical lighting, reduction of cooling load. *Renewable and Sustainable Energy Reviews, 41*, 1–13.

7. International Energy Agency. (2006). *Light's labour's lost policies for energy-efficient lighting*. Paris: IEA Head of Publications Service. Retrieved from https://www.iea.org/publications/freepublications/publication/light2006.pdf.
8. Energy Saving Trust. (2006). *Energy efficient lighting—Guidance for installers and specifiers*. CE61. London: Energy Saving Trust. Retrieved from http://www.rebelenergy.ie/ce61.pdf.
9. Pattison, P. M., Hansen, M., & Tsao, J. Y. (2009). LED lighting efficacy: Status and directions. *Comptes Rendus Physique, 41*, 285–286.
10. Loe, D. (2018). Energy efficiency in lighting—Considerations and possibilities. *Lighting Research and Technology, 19*, 134–145.
11. European Committee for Standardization. (2011). *Standard EN12464-1:2011 Light and lighting—Lighting of work places—Part 1: Indoor work places*. Brussels: Management Centre. Retrieved from https://www.cen.eu/Pages/default.aspx.
12. Galasiu, A. D., Newsham, G. R., Suvagau, C., & Sander, D. M. (2007). *Report NRCC-49498 Energy saving lighting control systems for open-plan offices: A field study*. Ottawa: Institute for Research in Construction, National Research Council of Canada. Retrieved from https://nparc.nrc-cnrc.gc.ca/eng/view/accepted/?id=b23dfd7d-3280-4740-b2fa-c57cd48806e9.
13. Haase, M., Skeie, K. S., & Woods, R. (2015). The key drivers for energy retrofitting of European shopping centres. *Energy Procedia, 78*(C), 2298–2303.
14. Chew, I., Karunatilaka, D., Tan, C. P., & Kalavally, V. (2017). Smart lighting: The way forward? Reviewing the past to shape the future. *Energy and Buildings, 149*, 180–191.
15. Dubois, M. C., & Blomsterberg, A. (2011). Energy saving potential and strategies for electric lighting in future North European, low energy office buildings: A literature review. *Energy and Buildings, 43*(10), 2572–2582.
16. Pandharipande, A., & Newsham, G. R. (2018). Lighting controls: Evolution and revolution. *Lighting Research & Technology, 50*, 115–128.
17. Yu, X., & Su, Y. H. (2015). Daylight availability assessment and its potential energy saving estimation—A literature review. *Renewable and Sustainable Energy Reviews, 52*, 494–503.
18. Rossi, M., Pandharipande, A., Caicedo, D., Schenato, L., & Cenedese, A. (2015). Personal lighting control with occupancy and daylight adaptation. *Energy and Buildings, 105*, 263–272.
19. Montoya, F. G., Pena-Garcia, A., Juaidi, A., & Manzano-Agugliaro, F. (2017). Indoor lighting techniques: An overview of evolution and new trends for energy saving. *Energy and Buildings, 140*, 50–60.
20. Gayral, B. (2017). LEDs for lighting: Basic physics and prospects for energy savings. *Comptes Rendus Physique, 18*, 453–461.
21. Ahn, B.-L., Jang, C.-Y., Leigh, S.-B., & Jeong, H. (2014). Analysis of the effect of artificial lighting on heating and cooling energy in commercial buildings. *Energy Procedia, 61*(Supplement C), 928–932.
22. Hee, W. J., Alghoul, M. A., Bakhtyar, B., Elayeb, O., Shameri, M. A., Alrubaih, M. S., & Sopian, K. (2015). The role of window glazing on daylighting and energy saving in buildings. *Renewable and Sustainable Energy Reviews, 42*, 323–343.
23. Shen, E., Hu, J., & Patel, M. (2014). Energy and visual comfort analysis of lighting and daylight control strategies. *Building and Environment, 78*, 155–170.
24. De Luca, F., Voll, H., & Thalfeldt, M. (2016). Horizontal or vertical? Windows' layout selection for shading devices optimization. *Management of Environmental Quality: An International Journal, 27*(6), 623–633.
25. De Luca, F., Simson, R., & Kurnitski, J. (2016). Energy and daylighting performance design of skylights and clerestories in a large hall retail building. In *Proc. Clima 2016—12th REHVA World Congress*, 22–25 May 2016 (Vol. 2). Aalborg University, Department of Civil Engineering, Aalborg, Denmark. Retrieved from http://vbn.aau.dk/en/publications/clima-2016%2D%2Dproceedings-of-the-12th-rehva-world-congress(f66ed00a-82ec-4f39-a4e9-f943ef3b44ee).html.

26. Estonian Government. (2012). *Methodology for calculating the energy performance of buildings* (Hoone energiatõhususe arvutamise metoodika). Passed 08.10.2012 Annex 63, Tallinn. Retrieved from https://www.riigiteataja.ee/en/eli/ee/MKM/reg/520102014002/consolide.
27. Reinhart, C. F., Mardaljevic, J., & Rogers, Z. (2006). Dynamic daylight performance metrics for sustainable building design. *Leukos, 3*(1), 7–31.
28. Reinhart, C. F. (2004). Lightswitch-2002: A model for manual and automated control of electric lighting and blinds. *Solar Energy, 77*(1), 15–28.

A Methodology to Address the Gap Between Calculated and Actual Energy Performance in Deep Renovations of Offices and Hotels

Robert Cohen and Greg Waring

Abstract A European Commission-funded Horizon 2020 project named ALDREN (ALliance for Deep RENovation in buildings) https://aldren.eu/ aims to establish the business case for deep renovation. The 30 month programme which started in November 2017 intends to encourage investment and accelerate the movement towards a nearly zero energy non-residential building stock across the EU, as targeted by 2050 to meet Paris Agreement commitments. The back-bone of ALDREN is the EVCS (European common Voluntary Certification Scheme) (Ribeiro serrenho T, Rivas Calvete S and Bertoldi P Cost-benefit analysis of the EVCS implementation, EUR - Scientific and Technical Research Reports, 2017) which will be used to track the deep renovation process. This paper describes the processes and tools being developed to close the gap between calculated and measured energy performance (EP):

1. A framework allowing measured (operational) performance to be compared with predicted (design) performance across all the countries in the ALDREN consortium using a harmonised approach and *common language* fed by a glossary of terms.
2. A "design for measurability" *protocol* that tracks the actions required during the deep renovation process, to ensure that performance predictions are as realistic as possible, that the construction and commissioning process is true to the design intent, and allowing the predicted performance to be verified through measurements.
3. A performance verification *tool*, which allows the predicted and actual (measured) performance to be compared at different levels of granularity.

The paper concludes that nearly zero energy performance targets can become measured outcomes, where driven by client leadership and wider team buy-in, and using the power of advanced simulation of HVAC systems to optimise design and ensure operation is aligned with the design intent.

R. Cohen (✉) · G. Waring
Verco Advisory Services Ltd., Corsham, UK
e-mail: Robert.cohen@vercoglobal.com

P. Bertoldi (ed.), *Improving Energy Efficiency in Commercial Buildings and Smart Communities*, Springer Proceedings in Energy,
https://doi.org/10.1007/978-3-030-31459-0_10

1 Background

A European Commission-funded Horizon 2020 project named ALDREN [1] aims to establish the business case for deep renovation. The back-bone of ALDREN is the European common Voluntary Certification Scheme (EVCS) [2] which will be used to set targets, track and verify the deep renovation process.

Two thirds of existing buildings in the EU are expected to be still in use in 2050, 30 years from today [3]. Many commentators assess that the EU's contribution towards the goals of the Paris Agreement can only be achieved if the energy demand from most of these buildings is drastically reduced by deep retrofits. Recital 16 of the Energy Efficiency Directive [4] defines "deep renovations" in a broad way, as *renovations which lead to a refurbishment that reduces both the delivered and the final energy consumption of a building by a significant percentage compared with the pre-renovation levels leading to a very high [efficient] energy performance*. One estimate is that only 1% of current renovations achieve this [5]. EuroACE believes that only around 15% of current building refurbishments incorporate *any* energy efficiency improvements, let alone a deep retrofit.[1] The Renovate Europe campaign proposes an energy demand reduction target of 80% by 2050 from 2005 levels [6].

The aim of ALDREN is to establish the business case for deep renovation in *non-residential* buildings with a focus on offices and hotels. The 30 month programme which started in November 2017 intends to encourage investment and accelerate the movement towards a nearly zero energy non-residential building stock across the EU, as targeted by 2050 to meet Paris Agreement commitments.

A key attribute of the approach to be adopted by ALDREN is the idea of energy performance verification. This means that the energy performance predicted at the design stage of a deep renovation will be verified by measurements during the first year of full occupancy. A protocol will be applied during the four critical stages of a deep renovation to ensure that performance outcomes achieve the design intent:

1. *Design Stage*:
   (a) Where possible, start with a detailed simulation model of the existing building and calibrate its predictions of energy performance by comparison with measured data.
   (b) Agree list of building improvements (fabric, plant, controls) using the calibrated model.
   (c) Ensure coherence between the detailed simulation model, control strategy, sub-meter plan, predicted energy budgets for sub-systems and each sub-meter and the EVCS target EP level.
2. *Construction*:
   (a) Ensure coherence of the as-built renovation with design intent and simulation model.
   (b) Revise energy budgets and EVCS EP level in response to any changes to design.

[1] Adrian Joyce, Secretary General, EuroACE, presentation to ALDREN partners meeting, Brussels, 10 April 2018.

3. *Commissioning and early occupation*:
   (a) Align the description of operations with both the simulation model and the as-constructed building and its HVAC systems and their controls. Rerun simulation model if actual operation of building (occupancy hours and density, weather, etc.) deviates significantly from the conditions of use assumed in the design stage predictions.
   (b) Confirm sub-meters and controls are installed, implemented and operate as intended.
4. *Operation*:
   (a) Fine tune the BMS controls to align with the description of operations.
   (b) Monitor monthly measured energy use against energy budgets for each sub-meter. Identify explanations for deviations from design intent and take remedial actions where appropriate.

This paper describes the methodology to be incorporated to support the verification of energy use.

2 Introduction

The primary functional purpose of an office building is to create a comfortable, healthy and conducive environment in which its occupants can carry out their work productively. Hotels have a slightly more complex purpose embodying "work, rest and play".

Climate change and other stresses on energy supply infrastructure dictate that office buildings and hotels should be energy efficient. It is a startling fact that we have no idea if they are: we do not attempt to predict it and we do not measure it. This is surprising given the widespread clamour for buildings to be energy efficient and their significant share of EU energy demand. Policy makers, investors and developers are starting to notice this lacuna in the struggle to mitigate climate change.

The thesis of ALDREN is that if we are going to start measuring the actual energy efficiency of buildings, then investors, developers and owners will want confidence that their renovated buildings will perform well. ALDREN is a process to underwrite the operational performance of building retrofits.

Many people assume that there are regulations which assure that deep retrofits produce energy efficient outcomes. Indeed, there are endless initiatives seeking to impose even stronger versions of these regulations, expecting as a consequence to secure even better energy efficiency outcomes, with some even demanding that these regulations mandate "nearly zero" energy buildings. But the awkward fact remains that unless performance outcomes are measured, no one can say whether these design-based regulations are effective in achieving their intended outcome.

Our readiness to admit that we have a broken system has had a sobering awakening through the knowledge that another country, with a commercial office market not dissimilar to those in the EU, has totally transformed the energy efficiency of its office buildings over the last 15 years. Australia has placed measured performance outcomes at the heart of its approach, but has simultaneously demanded a relentless focus on performance throughout the design, construction and early operation phases, starting with an operational performance target being set in stone by the developer from the outset.

Although the logic for an energy efficient solution is unassailable, it is a notoriously weak driver of institutional and behavioural change, especially when the evidence is invisible. Importantly, the case for ALDREN rests on more fundamental motivations. The paramount purpose of a commercial office building is to generate profits for the shareholders of the developer and investor businesses involved. The Australian market shows that office buildings with better energy efficiency ratings produce higher yields, through higher income returns and stronger capital growth. This is because better rated buildings are seen as better quality buildings, and command rent premiums tenants are willing to pay.

The pitch to tenants can also play to contemporary desires for new buildings to be smart and promote occupant well-being. What could be smarter than a building designed with advanced simulation? And what will deliver better well-being than building services operating efficiently and as intended?

ALDREN aims to learn from Australia's success in improving the energy efficiency of commercial office buildings. What has been achieved in Australia and how the UK compares (as an example of an EU country) are detailed in Appendix 1. The work on hotels will build on previous EC-funded research [7]. The ALDREN project plans to test the protocol on up to 15 real renovation pilot projects.

3 Performance Verification Protocol

The proposed process for ensuring the predicted energy performance of a deep retrofit is secured by verification using measured data is illustrated in Fig. 1. The process applies equally to offices and hotels; the same methodology will be used for both, although some differences in detail may emerge.

The key steps are as follows, with the numbering referring to Fig. 1:

1. Calculate asset rating for existing building under standard conditions.
2. Calculate asset rating for existing building under actual conditions.
3. Measure existing building energy use, compare with predicted energy use and calibrate model to match measurements.
4. Use calibrated model to agree list of building improvements for fabric, plant, controls, and so on.
5. Calculate asset rating under standard conditions for upgraded building after completion.
6. Calculate asset rating under actual conditions after year of operation.
7. Compare measured vs. calculated energy at monthly and sub-meter resolution.

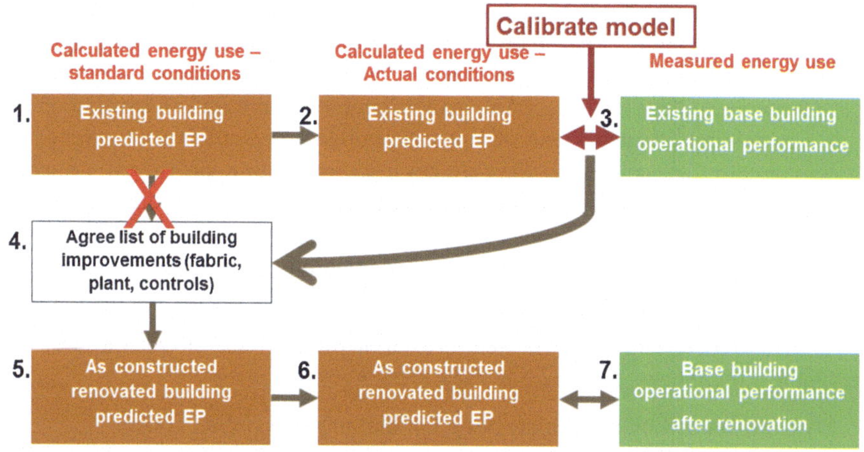

Fig. 1 Planned energy performance verification process for deep retrofits

Within steps 3 and 7, it is essential that comparisons of modelled and measured energy use are made on a like for-like basis:

1. Each is subject to the same boundary conditions. Typically this means ensuring that tenant hours of use, people and equipment density and weather over the year of measurement are applied in a re-run of the model.
2. Care is taken to compare results for the same energy uses. This requires a good understanding of what is being measured by each sub-meter and mapping this to energy metering points in the model.

It is apparent that the proposed process might be considered idealistic. For example, determining actual hours and density of use accurately could be time intensive for collecting reliable data, but the effort expended can be moderated according to the enthusiasm of stakeholders to check the match between design and actual, and learn from the deviations. There should be an expectation that metered values will be close to the simulated targets, but there are three fundamental challenges:

1. *Inadequacies of models*. Commonly used asset rating models are thought unlikely to be able to predict actual energy use by sub-meter accurately. ALDREN proposes that more advanced models are used, in which the building and its HVAC system and controls are modelled simultaneously and in explicit detail. Thereby simulating how the HVAC system would operate and be controlled to meet the predicted building heating and cooling zonal loads.
2. *Inefficient building operation*. It is known, regrettably, that many (if not most) large air-conditioned office buildings suffer from multiple imperfections in the way they are operated and controlled. Control systems are not designed and specified in sufficient detail, do not enable HVAC and lighting service levels to be tailored to demand, and rarely set out to limit services to unoccupied parts of a building (voids and when out of hours use is requested by some tenants, etc.).

BMS controls are often not working effectively for energy efficiency, being set to avoid complaints rather than *also* achieve efficient operation. Metering systems are often installed in buildings but are ignored or do not work. The building managers and their Facilities Management teams lack the skills needed to operate a building efficiently and do not have building performance requirements written into their contracts—they are not incentivised to improve performance. ALDREN proposes a lot more effort is made to ensure the real building operates as closely as possible to the model and design intent.
3. *Indoor environmental quality*. It is possible conditions in the real building will not match the idealised conditions assumed in the model.

The ALDREN energy performance verification protocol will incorporate the following five ingredients to overcome these challenges[2]:

1. Dynamic modelling to simulate design of building and HVAC system
2. Independent design review to confirm design and modelling are robust
3. Final design to include a validation plan (meter layout and targets) and preliminary Description of Operations
4. Tendering, construction and completion—Description of Operations revised
5. Diagnose and implement control tune-up and improvements

These steps are described in more detail below.

3.1 Dynamic Thermal Simulation of Building Design and HVAC System

Dynamic thermal simulation of the building and its HVAC system and controls, with a time step of 1 h or less, should be undertaken to predict heating and cooling demands under a range of expected conditions of use. The level of simulation proposed for ALDREN differs from current practices but is not ground-breaking, in the sense that it has become routine practice in Australia [8], and is used to some extent in the US under the guidance of ASHRAE 90.1 [9]. ASHRAE also offers an accreditation scheme for "Building Energy Modelling Professionals" the BEMP Certification [10]. BEMP certification validates competency to do the following:

1. Model new and existing buildings and HVAC systems with the full range of building physics.
2. Evaluate, select, use, calibrate and interpret the results of energy modelling software where applied to building and HVAC systems energy performance and economics.

[2] ALDREN will also be addressing indoor environmental quality, but this is not covered in the present paper.

Worldwide there are currently some 450 people with the BEMP qualification, of which some 350 are in North America, not surprisingly given its ASHRAE roots and that the exam uses imperial units. The EU probably needs to develop something comparable to BEMP, to give confidence to clients procuring advanced simulation services and to underpin the quality of advanced simulation.

It should also be noted here that in some respects the UK holds a world-leading position in advanced simulation alongside Australia and the US, by virtue of originating two of the most commonly used advanced simulation software platforms: EDSL-TAS Systems and IES Apache HVAC. Both these software packages are developed and operated by companies based in the UK, have full advanced simulation capability and are widely used by practitioners in the Australian market. Both are also used in the UK for HVAC simulation, but to a very limited extent, due to a lack of market demand. One of the most commonly used software platforms for detailed simulation of HVAC system in the US is EnergyPlus. Several commercial software packages are available which provide a user-friendly interface to the EnergyPlus engine, for example, DesignBuilder and Hevacomp.

There are several key objectives of advanced simulation:

- To understand how the HVAC system would operate for each hour of the year and thereby confirm plant capacity requirements more robustly. Load duration curves could be produced for each item of major equipment, enabling the designers to identify how much time would be spent in more or less efficient operating modes.
- To confirm that the proposed design is capable of meeting the base building energy performance target rating; typically the building should simulate to at least a quarter star better than target to engender confidence the actual operation will achieve the target.[3]
- To undertake "off-axis scenarios" to check the resilience of the base building rating to all plausible future scenarios for tenant hours and intensity of use and weather.[4] It is to be expected that different tenants may have longer or shorter hours of use than a base case standard condition, including late working, two-shift working and weekend working; some spaces may end up being unoccupied (void) for significant periods, and so on. A better rating will be achieved if the HVAC is designed so that different zones can be serviced independently and only occupied zones are serviced; otherwise the target might not be met.[5] A developer should expect the target rating to be achieved under all reasonable future scenarios.

[3] Base building energy ratings and the NABERS star rating scale are described in detail in Appendix 1.

[4] We note some MEP consultant engineers in Australia insist on minimum *tenant ratings* also being achieved (e.g. 1 star) before signing up to stretching base building targets, to give themselves perceived protection against excessive tenant energy intensity affecting the base building services efficiency. Modelling studies demonstrate this to be unnecessary (although truly agile working is sweating the system), but the ease of mind it affords is understandable.

[5] The NABERS base building rating defines energy efficiency using the principle that a building should receive no benchmark "allowance" from lettable space for any period it is unlet.

- To inform the development of a verification plan which identifies necessary sub-metering.
- To produce monthly targets for each sub-meter and each sub-system (heating, hot water, cooling, fans, etc.).
- To inform the optimisation of HVAC control.

3.2 *Independent Design Review*

An independent design review (IDR) should be undertaken by an independent and experienced energy efficiency professional who has been assessed for high levels of expertise in relation to:

- Deep retrofit building projects and the design of HVAC services and their controls
- Commissioning/tuning of buildings
- Energy auditing and energy efficiency improvement of buildings
- Simulation of building performance

The IDR scrutinises the design, metering plan and the simulation outputs with the overarching objectives of checking whether it is probable the building will achieve its target base building rating, identifying potential improvements in either the current design or the design team's next design or both, and generally disseminating good practice to the industry.

The typical output from an IDR would be a report in spreadsheet format that includes the following components:

- A review of each building services package, including mechanical services, electrical services (including lighting), hydraulic services and vertical transport; this will include commentary on:
  - Risks in design, construction and operation with consideration to the target energy performance level, environmental impact and maintenance.
  - Options, alternatives and avenues of enquiry that may assist the improvement of the design and effectiveness of controls.
  - Items within the design that may lead to shortcomings with regards to energy efficiency outcomes, environmental performance and/or maintenance requirements.
- A review of the proposed energy metering in order to provide commentary regarding the suitability and/or adaptation of the metering to measure post-construction outcomes.
- A review of the architectural design, considering layout, orientation, materials selection, glazing and shading.
- Detailed review of proposed control and/or the recommendation of optimised control approaches suitable for the project.

- Issues and recommendations relating to the proposed commissioning process and ongoing management practices, to help ensure that the building performs to its potential.
- A peer review of any already completed simulation work.

The detailed design review report will provide clear identification of issues along with specific recommendations for consideration and learning. It will also note issues that might be more appropriately considered for the next comparable design.

At least one workshop would be expected, in order to present the findings of the review to the design team and developer representatives.

The exact timing of an IDR can vary according to project circumstances, with earlier reviews giving more opportunity for changes before things become fixed, whilst reviews at a later stage can have the advantage of looking at a more developed design and, as a result, more meaningful modelling.

Producing recommendations in a spreadsheet format facilitates commentary by the design team and then the reviewer's responses to the design team's feedback and comments.

3.3 Final Design

The final design should consolidate any changes arising from the IDR into the design package. Key aspects relating to the performance target include a performance validation plan and the first draft of the proposed Description of Operations. It should be explicit at this stage how the sub-system monthly energy use predictions of the simulation model will be tracked and verified by measurements with sub-meters.

3.4 Tendering, Construction, Commissioning and Completion

During the tendering and construction stages, it is important to keep the simulation model and Description of Operations up to date with any significant design changes. For example, if any changes threaten the achievement of the target rating after a value engineering process, further modelling may be needed to demonstrate that the target would not be compromised.

The draft Description of Operations should be made available to tenderers for the controls engineering and used as an input into the design of the control system. The objective should be for the implemented control strategy to mirror the control system assumed by the simulation model. If the actual controls replicate how the building operates in the model, the actual HVAC system performance should be close to the simulated performance, giving confidence the target rating will be achieved. Any refinements introduced to the control strategy should be reflected in a revised version of the Description of Operations document that emerges on completion and handover.

3.5 Early Operation Intensive Fine Tuning

A key objective of commissioning should be to ensure the control algorithms in the completed building are consistent with the simulation model of the final design and the revised Description of Operations.

Nevertheless, for successful delivery of target ratings, monitoring and tuning during the year of the Defects Liability Period has been found to be essential. This typically includes:

- Establishment of building and subsystem targets based on the simulation.
- Tracking sub-metered performance measurements against predicted targets to produce monthly monitoring reports. Where significant discrepancies occur, suggestions should be made for their potential causes and the remedial actions which might mitigate them, and their feasibility either on the current project or for future projects.
- At least four BMS tuning exercises during the course of the defects liability period, each including a detailed review of operation.

QUANTUM Performance Test Bench

ALDREN partners are aware that in much of the EU, building services engineering teams currently lack the skills to do robust detailed simulation of HVAC systems and their controls, as it is done in Australia and North America. The ALDREN performance verification protocol will therefore offer an alternative approach which is being promoted by another H2020 project, QUANTUM [11]. The QUANTUM "performance test bench" tool aims to get written into tender documents (and so the Contract) every significant data point for plant operation and control (this could be up to 150 data points for a large complex air-conditioned office). The contractor is then required to define what should happen for this data point when the building is operating (e.g. as an illustration, a fan motor speed when the building zone it is supplying is 50% occupied), and the means to test if that is happening correctly.

Once the building is in early operation, an independent verifier is asked to compare what is happening for that data point with what the contractor said should be. This produces a robust defects list for the contractor to resolve, so that the building operates as intended.

4 Monitoring and Verification

In order to verify the predicted energy performance of a renovated building in operation, the ALDREN project will develop a tool (in Microsoft Excel®) which acts as a repository for the data generated by the various protocol stages described above, and presents a side-by-side comparison of energy consumption predictions and measured data.

The tool will also collect relevant contextual data which describe the actual conditions in a building during the assessment period and can be used to tailor a reference value or benchmark that applies to standard conditions of use and that might be deployed to produce an asset or operational rating.

Development of a harmonised European procedure to calculate an operational rating, for example on an A to G or number of stars scale, falls outside the scope of the ALDREN project. However, it is proposed to include within the ALDREN process collection of the relevant contextual data (e.g. occupied hours and a schedule of activities by functional space in and around the building) to enable the ALDREN pilot projects to produce operational ratings using existing schemes (e.g. the Landlord Energy Rating) [12].

The structure of the performance verification tool (PVT) is presented in Fig. 2 (this relates to a specific retrofit stage, e.g. steps 1, 2 and 3 or 5, 6 and 7 in Fig. 1).

As indicated in the flowchart, the tool will collect data describing the predicted energy use of the building for regulated energy loads under standard conditions and actual conditions, the measured energy consumption, and a range of building characteristics data which are relevant to the simulation process and calculation of operational ratings.

The tool will be designed to accept data from models at varying levels of simulation complexity expected in different retrofit projects.

- At a "basic" level, the simulation would be expected to produce outputs at energy end use granularity and with a monthly breakdown.
- Where "advanced" simulation is applied, the calculation would be expected to produce monthly energy consumption predictions for each sub-meter that is material to regulated energy loads.

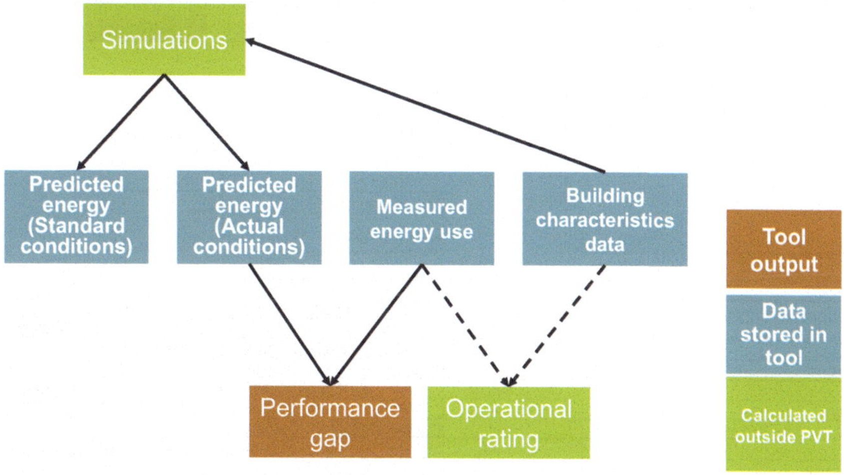

Fig. 2 Outline of the performance verification tool at a specific retrofit stage

The output for a “basic” assessment is proposed to highlight the variance for each end use, using a red-amber-green rating applied to the degree of variance, allowing design teams to prioritise areas with large relative or absolute variance for further investigation. A purely illustrative summary of the proposed style of annual assessment by energy end use is presented in Table 1. An equivalent output at monthly resolution will permit an understanding of seasonal variances in each energy end use.

Where an advanced simulation is employed, a more detailed output is proposed, with meter level granularity on a monthly basis, and allowing a user adjustable tolerance on variance, based on either a percentage or absolute value. A purely indicative example is illustrated in Table 2—the values in both Tables 1 and 2 have been made up to illustrate the method—they do not represent data from a real building.

It is recognised that a key challenge will be the matching of the physical energy metering in retrofitted buildings to the energy scope used for the assessment. Current design procedures do not routinely deliver discrete metering of each regulated energy end use in buildings (despite this being required in building regulations in

Table 1 Indicative output for annual energy kWh values by end use from a “basic” assessment: predicted under actual conditions, measured and the variance between these

End use	EVCS-actual conditions	Measured	Variance (kWh)	Variance (%)
Space Heating	8,775,673	5,072,482	-3,703,191	-73%
Hot water	674,887	826,126	151,240	18%
Refrigeration	2,136,180	2,340,840	204,660	9%
Fans	121,605	102,209	-19,397	-19%
Pumps	1,430	1,421	-9	-1%
Controls	1,128,920	980,400	-148,521	-15%
Humidification	1,007,449	1,073,800	66,351	6%
Lighting (internal)	774,040	822,385	48,345	6%
Total	14,620,184	11,219,664	-3,400,521	-30%

Table 2 Illustrative values for the variance in monthly energy kWh values between predictions by an advanced simulation and actual measured consumptions, at the level of each meter

Select tolerance method:	By % of meter

	High	Low
By consumption (kWh)	50,000	10,000
By % of meter	25%	10%

Meter	Meter type	Energy coverage	Jan-17	Feb-17	Mar-17	Apr-17	May-17	Jun-17
Main electricity meter	Main meter	Whole building	-94,021	101,480	52,730	-15,945	48,321	44,997
Chillers	Sub-meter	Whole building	-85,081	-72,125	-53,667	-63,361	-86,253	-133,667
AHU1	Sub-meter	Whole building	1,513	2,271	2,469	1,903	2,572	1,535
AHU2	Sub-meter	Whole building	32,850	54,714	9,047	-15,994	3,400	-69,656
Mech panel	Sub-meter	Whole building	885	378	247	762	679	1,112
Lifts	Sub-meter	Whole building	-155	-282	16	87	121	214
Floor 5 packaged AHU	Sub-meter	Tenant	2,378	2,410	467	-2,353	1,372	-2,468
Lighting - common parts	Sub-meter	Landlord	-215	-307	-464	-362	-443	-522
Lighting - Tenant	Sub-meter	Tenant	-120	-15	295	435	307	749
Plug loads - common parts	Sub-meter	Landlord	-12,038	-15,038	-4,347	-9,512	-11,416	-14,128
Plug loads - tenants	Sub-meter	Tenant	-57,922	-32,352	218	3,999	42,894	86,117

some member states). The ALDREN protocol will identify the actions required in order to deliver this metering in practice, but it is recognised that some manipulation of data may be required outside the verification tool in order to achieve a like-for-like comparison.

5 Conclusions

ALDREN is a new H2020 project focusing on deep retrofits, that includes the objective to offer energy performance verification by measurements of energy use in operation.

A methodology has been proposed for how this will be done. It centres on ensuring the verification process is based on like-for-like comparisons of predictions and measurements.

The procedure allows for using more or less detailed models for the performance predictions, and steps to ensure a building is operated according to the design intent.

It is anticipated that using detailed simulation of HVAC systems and their controls, alongside the dynamic thermal simulation of the building itself, for prediction and target setting, will enable performance outcomes to come close to matching design aspirations in air-conditioned buildings.

The proposed methodology will be tested during the ALDREN project on up to 15 pilot study buildings spread around the countries where the ALDREN partners are based (France, the UK, Belgium, Spain, Denmark and Slovenia) and possibly also in countries where the QUANTUM project is being tested (Germany, Austria and Switzerland).

Acknowledgements The ALDREN project has received funding from the European Union's Horizon 2020 research and innovation programme under grant agreement No 754159.

Appendix 1: What Has Been Achieved in Australia?

Some 15 years ago in Australia, "base building" energy ratings[6] had started to influence investment decisions for existing and new buildings, sales and purchases. The scheme that measured and verified this base building performance was called the

---

[6] Base building energy covers the following energy end uses; sub-meters should be provided to measure the energy consumed by fuel type in supplying each of these building central services:

- Heating, domestic hot water, cooling and ventilation, for example, to a BCO Guide specification*.
- Common-area lighting and power (including lift lobbies, plant rooms and common-area toilets).
- Vertical transportation (e.g. lifts and escalators).

National Australian Built Environment Rating System or NABERS [13]. Some of the key steps have been:

- 1999: New South Wales introduced a voluntary system (the Australian Building Greenhouse Rating, ABGR), to measure and benchmark the energy use of existing office buildings. This developed into the NABERS national Scheme [14].
- 2002: Commitment Agreements were conceived for developers to ensure new offices could operate at their target energy performance levels and enable occupiers to sign up to pre-lets for space with the in-use energy performance they wanted.
- 2004: State governments started to set minimum standards for space they occupied. New South Wales took the lead in March 2004, when they decreed their existing owned buildings and tenancies had to be rated by the year end, should attain 3 star base building (on a 1 to 5 star scale, with the empirical median performance at 2.5 stars) by July 2006 and new leases should require 3.5 stars from 2006 [15]. They also required 4 stars for major upgrades and 4.5 stars for new buildings. Other States gradually introduced their own minimum standards.
- 2006: the Federal Commonwealth (Australian) Government mandated 4.5 star base buildings for new buildings, major refurbishments and new leases over 2000 m^2. Most States have since ratcheted up their requirements to the 4.5 star level for all their stock over 2000 m^2. In the same year, the Property Council of Australia introduced minimum NABERS base building energy ratings into their definitions of new offices: 4.5 stars for grade A, 4 stars for grade B.
- 2010: the federal government introduced the Building Energy Efficiency Disclosure Act, to mandate disclosure of Base Building ratings on sale or let of office premises over 2000 m^2 NLA.
- 2011: NABERS extended the top of their scale to 6 stars, stating 5 stars represented excellent performance, and 6 stars market leading [16]. The new 6 star level was set by taking a theoretical 7 star level as zero emissions and applying a 50% reduction in the emissions at 5 stars. Similarly, 5.5 stars is a 25% reduction from the 5 star level.
- 2012: the energy performance bar for grade A offices was raised: to 5 stars for new buildings and to at least 4 stars for existing buildings [17].
- 2017: the threshold for mandatory disclosure was reduced from 2000 to 1000 m^2 NLA [18].

A feasibility study [19] published by the Better Buildings Partnership (BBP) in May 2016 confirmed that, in the commercial office property market in Australia,

- Exterior lighting.
- Exterior signage provided by the building owner for the benefit of office occupiers.
- Generator fuel where it serves central services.
- Car park ventilation and lighting, where internal or external car parks within the legal boundaries of the site are provided for occupier use.

*Supplementary HVAC services to a tenant's energy-intensive areas including server rooms, dealer rooms and laboratories should use energy off the tenant's meter, not the landlord's HVAC.

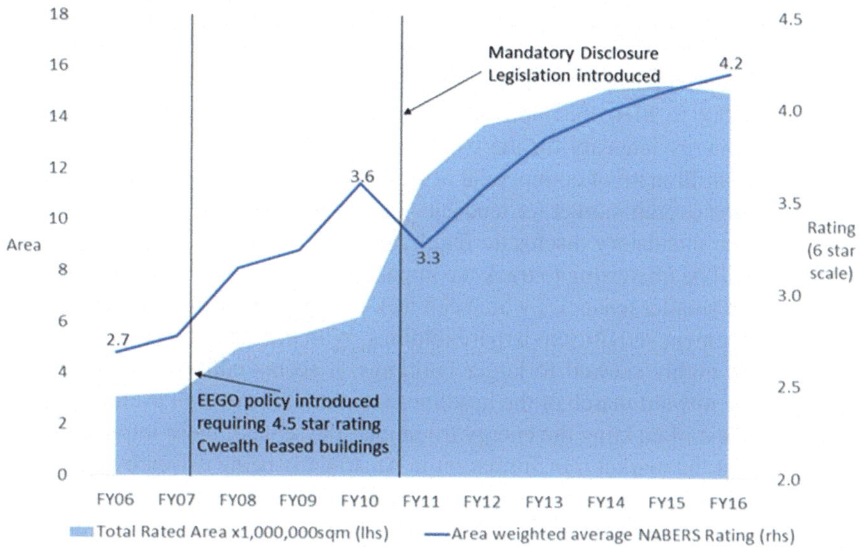

Fig. 3 Growth in rated commercial office floor area and improvement in the existing stock average base building energy rating from 2006 to 2016. (Source: NABERS, OEH [20])

better base building operational energy performance has become aligned with investor, developer and occupier interests. Over the last 15 years, this has driven a systemic change in design, construction and operation of office buildings, with innovation flourishing across the supply chain. As a result, base building services in today's *new* buildings in Australia use on average half the energy they did when measurements started in 1998, and the best one fifth. The nexus of financial and property industry interests has also driven a remarkable uplift in the base building energy performance of the *existing* stock in Australia (see Fig. 3).

The context for Fig. 3 is that when base building ratings were initiated in 1998 (on a voluntary basis), a scale from 1 to 5 stars was set using empirical data to position the average performance at 2.5 stars. The blue line in Fig. 3 shows that by 2006 the average rating had crept up to 2.7 stars (right hand scale). By then some 3 million m^2 of commercial office floor space had a rating (the blue-filled area on the graph and left hand scale).

Progress was boosted in 2007 by an Energy Efficiency in Government Offices policy requiring office buildings leased by the Commonwealth government to be a minimum of 4.5 stars. By 2010, the average rating had climbed to 3.6 stars, a 24% improvement on the 2006 position.

In 2010, policy makers had sufficient confidence in the approach to mandate disclosure for sale or let transactions [21], which widened the empirical data from a voluntary cohort of buildings. Not surprisingly, the overall effect was a reduction in the average rating over the next year to 3.3 stars as poorer performing buildings were obliged to lodge their rating data, as well as those that had been doing so voluntarily.

However, the hiatus in average rating improvement was short-lived and within a couple of years the market average was exceeding its previous record high and indeed grew continuously every year, reaching 4.2 stars by 2016. Over the ten year period from 2006 to 2016, the improvement from 2.7 to 4.2 stars represented a 41% reduction in energy intensity for the whole of the rated stock, which by then had climbed to 16 million m^2 of commercial office floor space, an almost complete penetration into the overall market for tenancies over 2000 m^2.

In 2017, the mandatory disclosure requirement was extended to tenancies over 1000 m^2. It will be interesting to track the impact on the stock average rating once it includes these smaller tenancies which can lack the economies of scale supporting energy management activities in larger buildings. With the allocation of floor space in the market highly skewed to larger buildings, it seems unlikely this step will undermine the upward march of the headline statistic for the overall average rating.

In the context of tackling the energy trilemma,[7] the scale of these improvements is striking. But the market transformation in Australia is being driven by commercial interest: investors and developers get better yields from better rated buildings because occupiers associate them with better buildings.[8] Statistics demonstrate that occupiers stay in better rated buildings longer—voids are lower—as shown in Fig. 4a [22]. Occupiers are also willing to pay higher rents for better rated buildings, so income return is higher, as shown in the left hand set of data in Fig. 4b [23]. Better rated buildings also produce stronger capital growth (middle set of data in Fig. 4b).

Government's role has been to develop and operate an online public rating and disclosure platform, create infrastructure for independent and authoritative ratings to be produced by accredited assessors and to lead by example by setting minimum ratings for the space it leases. Once the rating had become established in the market,

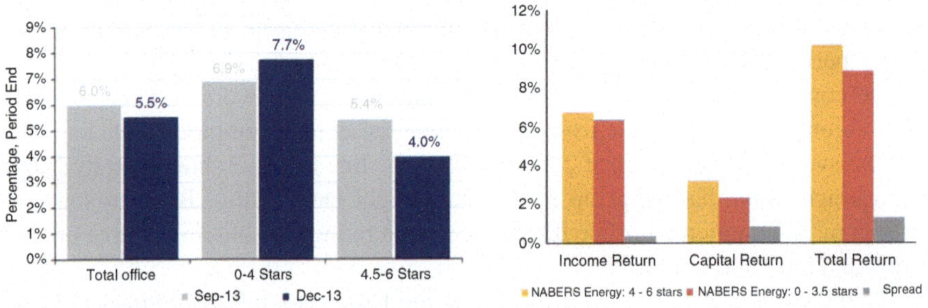

Fig. 4 (**a**) Offices with higher NABERS Energy ratings have lower voids (*y*-axis shows % vacancy rate at end of given period). (Source: NABERS, IPD). (**b**) Offices with higher NABERS Energy ratings deliver stronger financial returns (*y*-axis shows % financial return). (Source: The Property Council/IPD Green Property Index, MSCI, March 2015)

[7] Climate change, security of supply and affordability (minimising energy costs).

[8] The underlying logic is that a better rating is associated with a building that has been better designed, better constructed, better commissioned and better operated and maintained.

government was moved to make performance disclosure mandatory. It is apparent that technical innovation usually needs policy intervention to extend market take-up beyond early adopters. But the experience in Australia demonstrates how performance transparency can be powerful in driving improvement, both at the top and the bottom of the efficiency scale: there are no mandated minimum energy standards.

Appendix 2: How Does the UK Compare?

By contrast with Australia, sale and let transactions in the UK are informed by an Energy Performance Certificate (EPC) [24], a theoretical calculation which rates how energy efficient your building is using grades from A to G (with "A" the most efficient grade) but does not reflect real performance and so gives limited insight to decision makers. Full compliance with Building Regulations Part L2A [25] does support a direction of travel which should make it possible to *measure* the performance outcomes for all the energy uses regulated by Part L2—using sub-metering which has been mandated for new buildings since 2002 [26]. However, there is no requirement, nor a pervading culture, for a comparison to be made between the measured outcomes and the predictions made at the design stage, let alone for this to be disclosed to stakeholders. In many respects, this is a perverse situation—why is there no guidance suggesting this would be a useful purpose for the metering system? The absence of such a culture (or requirement) means that this comparison is almost never made. It certainly prevents policy makers getting the evidence base for the ratcheting up of Part L2 requirements that has occurred roughly every 5 years since energy efficiency regulations were first introduced for commercial buildings (offices and shops) in 1985 [27]. And it prevents a light being shone on the notorious performance gap between the predicted and measured values for regulated energy end uses.

This failure to use evidence which could be collected from equipment installed to comply with regulations to tackle the performance gap is especially stark in a building with a single occupier, where all sub-meter data can reasonably be expected to be collected at a single central point.

In multi-let buildings, the Part L2 metering requirements are less well aligned with the objective of quantifying the energy performance gap. Individual tenants might not install their own sub-metering system for their own energy use. But even if they did, this data on energy end use breakdown would not normally be made available to a landlord, making it difficult to aggregate whole building energy use for each category of regulated loads and creating a barrier for making a comparison with the predictions at the design stage.

Unlike in Australia, the UK does not have a mentality of designing for measurability. The best empirical evidence available for commercial multi-let buildings is collected by the BBP from its members. This data enables a comparison to be made between metered whole building energy intensity and the building's EPC grade (see Fig. 5). The data suggests a limited correlation—the median energy intensity values do get better (lower) as grade improves, but there's so much variability that this marginal trend is of limited statistical significance.

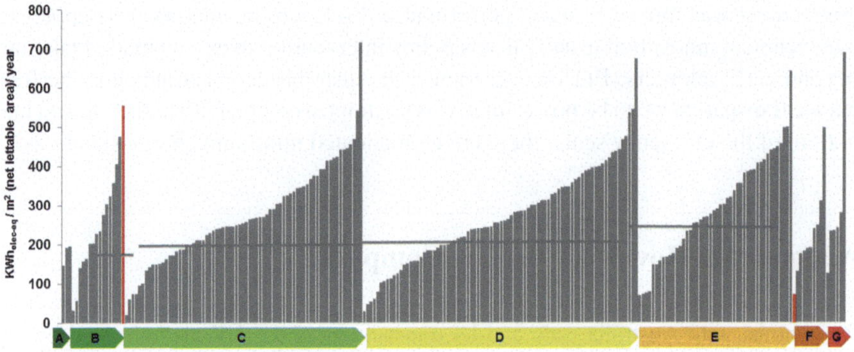

Fig. 5 Comparing whole building energy intensity for buildings with different EPC grades. (Source: Sarah Ratcliffe (BBP) at the CIBSE Building Performance Conference, London 17 November 2016 [28])

The UK's non-residential building construction supply industry is notoriously fragmented, a position often cited for poor energy performance outcomes. It is true that responsibility for energy efficiency is often passed from the building's MEP designers to the appointed Design and Build contractor. And once the shell-and-core is completed, often with a placeholder Category A fit-out, it is then handed over again to a whole new set of businesses to deliver the desired Category B fit-out for each tenant. However, new building procurement in the Australian market is not materially different in these respects, and yet because the energy performance outcome is a critical KPI for the developer in Australia, the baton is not dropped at each handover point in the energy efficiency 'relay'.

It is material to note that both jurisdictions share the aim to provide the market with relevant information about the energy performance of a building at the moment of a property transaction, when the data can inform buying and letting decisions. The arrangements in the UK arose from the implementation of the European Energy Performance of Buildings Directive [29], whilst those in Australia evolved from experience of applying the initially voluntary NABERS scheme. However, their approaches to the same end could not be more different: the UK's EPC [30] and the Australian Building Energy Efficiency Certificate (BEEC) [31]—see Fig. 6.

The alternative representations of a building's energy efficiency in each country for the purpose of market transparency (theoretical "asset rating" EPC vs. measured "operational rating" BEEC), gives rise to the idea of considering the different approaches as if they were medicines being tested in a medical blind trial to treat a disease. After at least 10 years of each jurisdiction applying their different "medicine", how have the two respective cohorts of patients (buildings) responded to the treatment they received. The 2016 feasibility study published by the BBP delved into the data to determine what, if any, differences there were in outcomes in the UK and Australia.

To make the comparison on a like-for-like basis, the energy performance of buildings in London and Melbourne were plotted on the same graph, where the

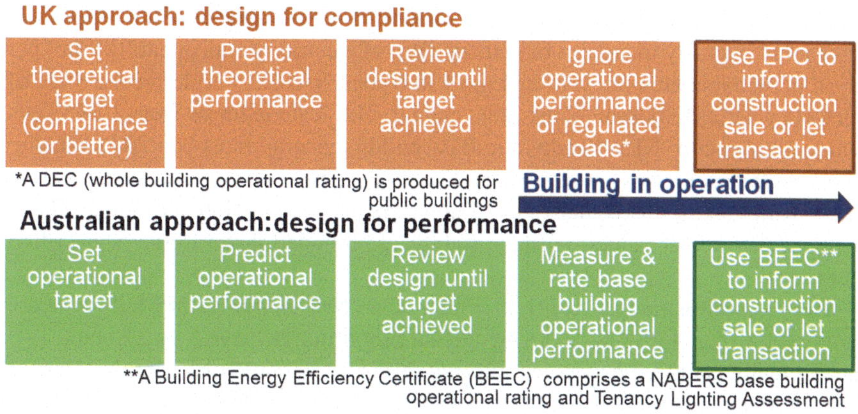

Fig. 6 Comparing how the markets in the UK and Australia are informed about building energy performance at the moment of a sale or lease property transaction

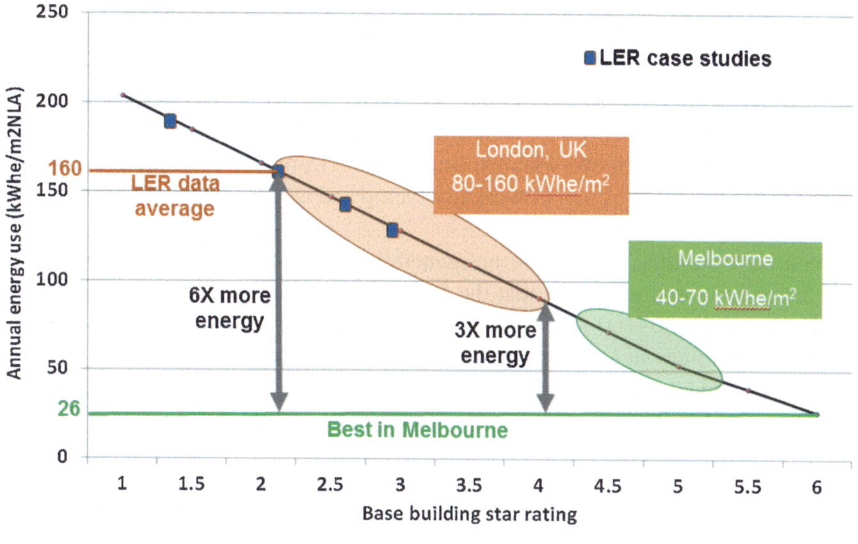

Fig. 7 Base building performance of new offices in Melbourne and London compared

x-axis is the NABERS 1 to 6 star scale and the *y*-axis is the measured base building energy intensity (see Fig. 7). Although there are significant differences between Melbourne's climate and London's, this factor would not be enough to drive dramatic variances in annual energy intensity. For much of a typical year in each climate, the weather in London and Melbourne is similar. Melbourne tends to have much hotter peak summer months, requiring *more* cooling energy, but this is compensated by milder peak winter months, requiring *less* heating energy.

The black line on the graph in Fig. 7 shows the relationship between base building energy intensity measured in units of kWh of electricity equivalent[9] (kWhe) per m^2 of net lettable area per year and the 1 to 6 stars NABERS scale for the State of Victoria where Melbourne is the State capital. The scale is linear from 1 to 5 stars with a 38 kWhe/m^2 NLA bandwidth. Base building energy must be <204 kWhe/m^2 to get on the scale with a 1 star rating. 5 stars is at 52 kWhe/m^2 NLA. 6 stars is at 26 kWhe/m^2 NLA, that is, half-way from 5 stars to net zero. Half stars are available between the integer values—official ratings are rounded down to the nearest half star rating.

The graph illustrates a reality in which new office buildings in Melbourne are never worse than 4.5 stars, and a significant proportion achieve 5 or 5.5 stars. Two have actually achieved 6 stars, confirming this level as market leading. In terms of base building energy intensity, this places new buildings in Melbourne in the range 40–70 kWhe/m^2/year, with the best at 26 kWhe/m^2/year.

The average base building energy intensity of 160 kWhe/m^2/year for London offices shown in Fig. 7 covers data collected for 85 assets by Verco in 2013. Because base building energy use is not specifically measured, the quality of this bulk data was recognised to be weak. To address this concern, detailed energy audits were undertaken for four of these assets, with the findings written up as case studies [32]. The results for these case studies were scattered around the average level, giving confidence in the average value, which was also anecdotally corroborated as plausible by individuals with everyday exposure to data from office building portfolios. This exercise was done as part of work to develop and test a Landlord Energy Rating (LER) scheme for the BBP.[10]

With base building energy use not generally measured in the UK, it was estimated, probably optimistically, that the energy intensity range for new buildings in London might be 80–160 kWhe/m^2/year. The conclusion is that the most efficient new office buildings in London are three times more energy intensive than the best in Melbourne, whilst the least good in London are using over six times more energy than the best in Melbourne. With no visibility of actual base building performance

[9] To calculate the kWh of "electricity equivalent" of total energy use, kWh of electricity are added to kWh of any fuel multiplied by 0.4 and kWh of hot or chilled water delivered to the building multiplied by 0.5. The kWhe metric enables timeless, international comparisons of a building's energy performance, and facilitates intrinsic building energy efficiency to be rated, independently from local, regional or national grid factors. Furthermore, with electricity often/usually the dominant energy carrier for commercial offices, kWhe has the enormous merit of using unity as the weighting or intensity factor for electricity—thus a unit of electricity retains the same value independent of the building's location around the globe, or the timing of the period for which the analysis is being undertaken.

[10] In 2012, BBP commissioned Verco and the UBT to develop the LER, a NABERS-style energy rating scheme for UK offices. Its application on about 85 buildings exposed challenges with the configuration and sub-metering of *existing* building services systems. This led BBP to focus on the concept of base building performance agreements for *new* buildings, where it was potentially possible to design out the obstacles to a harmonised investment-grade rating presented by the variability of engineering services and sub-metering configurations encountered in the existing stock.

outcomes, it is no coincidence that the base building energy efficiency of new UK commercial offices compares so unfavourably with that for their counterparts in Australia. Harking back to the medical trial, the patients in Australia have fared very well, whilst those in the UK remain critically ill.

What proves that climate is not the critical factor is the data for base building energy intensity for Melbourne offices in 2002. Back then, the average rating was 2.5 stars or about 150 kWhe/m^2/year. It can thus be seen that the least good new buildings in Melbourne are now using less than half the average in 2002, whilst the best are using six times less. The differences in today's outcomes between London and Melbourne are clearly being driven by the huge improvements that Melbourne has achieved in the last 15 years, not climate differences. The EPC has not driven corresponding improvements in the operational energy performance of the UK's commercial office buildings. However, Australia's experience suggests that with the right drivers, the energy use of base building services in *new* UK offices could typically be halved, and best practice four to five times lower.

Following reports on the performance gap by the Green Construction Board in 2013 [33] and UK Green Building Council in 2016 [34], the UK property market has woken up to the potential of buildings which perform as intended and to the risks with those that do not. The ability to demonstrate that energy efficient operation can be achieved in new buildings, can help to identify exemplar pathways for deep retrofits of the existing stock, on a trajectory towards net zero energy in operation.

References

1. ALDREN (ALliance for Deep RENovation in buildings). Retrieved from https://aldren.eu/.
2. Ribeiro Serrenho, T., Rivas Calvete, S., & Bertoldi, P. (2017). *Cost-benefit analysis of the Voluntary Common European Union Certification Scheme (EVCS) implementation*. EUR - Scientific and Technical Research Reports.
3. European Parliament. (2016). *Energy efficiency of buildings. A nearly zero energy future?* http://www.europarl.europa.eu/thinktank/en/document.html?reference=EPRS_BRI(2016)582022
4. European Union. (2012). Directive 2012/27/EU of the European Parliament and of the Council on Energy Efficiency. *Official Journal of the European Union, 315*, 1–56.
5. The Economist. (2013). *Investing in energy efficiency in Europe's buildings. A view from the construction and real estate sectors*.
6. The Renovate Europe campaign. Retrieved from https://www.renovate-europe.eu/.
7. Tsoutsos, T., Tournaki, S., & Frangou, M. (2016). *Nearly zero energy hotels in Europe (NEZEH)*. Technical University of Crete. Retrieved from www.nezeh.eu.
8. New South Wales Office of Environment and Heritage. (2019, February). *Handbook for estimating NABERS ratings*. Version 1.1. Hurstville: New South Wales Office of Environment and Heritage
9. ASHRAE Standard 90.1. (2016). *Energy standard for buildings except low-rise residential buildings*. https://www.ashrae.org/technical-resources/bookstore/standard-90-1.
10. U.S. affiliate of the International Building Performance Simulation Association (IBPSA-USA) and the Illuminating Engineering Society (IES). BEMP – Building Energy Modelling Professional Certification. https://www.ashrae.org/professional-development/ashrae-certification/certification-types/bemp-building-energy-modeling-professional-certification

11. Mechanical and Electrical Engineering Working Party of National, Regional and Local Authorities (AMEV) for QUANTUM project. (2017). *Technical monitoring as an instrument for quality assurance.*
12. Better Buildings Partnership (BBP). (2014, June). *Landlord energy rating documentation.* Better Buildings Partnership (BBP)
13. National Australian Built Environment Rating System (NABERS). Retrieved from https://www.nabers.gov.au/.
14. Australian Building Greenhouse Rating (ABGR). Retrieved from https://www.nabers.gov.au/about/our-story.
15. New South Wales State government decree. (2004). M2004-04 Greenhouse Performance of Government Office Buildings and Rental Properties.
16. Blundell, L. (2011, August). NABERS Energy goes to 6 stars as (most of) the industry moves on. *The Fifth Estate.*
17. NABERS. (2016, April). *Supporting policies and initiatives.* Version 0.6.
18. NABERS. (2018, September 30). *Annual report 2017–18* (p. 2). Version 1.
19. Bordass, W. T., Cohen, R. R., & Bannister, P. (2016). *Design for performance: UK Commitment Agreements: Making measured energy in-use the objective for new office buildings.* Feasibility Study Final Report. Better Buildings Partnership (BBP).
20. Harrison, S. (2017, June). *NABERS and mandatory disclosure.* Presentation to UCL/CIBSE NABERS workshop, London.
21. Australian Government. (2010). Building Energy Efficiency Disclosure Act No. 67 2010, to promote the disclosure of information about the energy efficiency of buildings, and for related purposes.
22. IPD/Department of Industry. (2013, December). *NABERS Energy Office Market Analysis, Figure 16.* IPD/Department of Industry.
23. The Property Council/IPD Australia, Green Property Index (2015, March). https://www.msci.com/documents/1296102/1672377/MSCI_AU+Green+snapshot+Flyer.pdf/e2548b3f-6809-4732-bd4d-281483e81256
24. UK government, Energy Performance Certificates for your business premises. Retrieved from https://www.gov.uk/energy-performance-certificate-commercial-property.
25. HM Government. (2006). *The Building Regulations 2000 Approved Document L2A: Conservation of fuel and power in new buildings other than dwellings* (2006 Edition). London: HM Government.
26. HM Government. (2002). *The Building Regulations 2000 Approved Document L2: Conservation of fuel and power in buildings other than dwellings* (2002 Edition). London: HM Government.
27. HM Government. (1984). *Building Act 1984.* Retrieved from http://www.legislation.gov.uk/ukpga/1984/55.
28. Ratcliffe, S. (2016, November). *Performance targeting & measurement is essential for effective energy efficiency policy & driving market transformation.* London: CIBSE Building Performance Conference.
29. ODPM. (2004). *Proposals for amending Part L of the Building Regulations and Implementing the EPBD.* Consultation Document.
30. HM Government. (2007). *The energy performance of buildings (certificates and inspections) (England and Wales) regulations 2007.* London: HM Government.
31. Australian Government. *What is a BEEC?* Retrieved from http://cbd.gov.au/get-and-use-a-rating/what-is-a-beec.
32. Waring, G., & Bordass, W. T. (2013, September). *LER Phase 2: Case studies.* Report to the Better Buildings Partnership.
33. Austin, B. (2013). *The performance gap—Causes and solutions.* Green Construction Board—Buildings Working Group.
34. UK Green Building Council (UKGBC). (2016, May). *Delivering building performance.*

Financing Models for Energy Efficiency in Public Buildings and Street Lighting in Germany and Neighbouring Countries

Aleksandra Novikova, Kateryna Stelmakh, Alexander Klinge, Ingmar Juergens, and Matthias Hessling

Abstract Investment in the upgrade of public infrastructure such as public buildings and street lighting provides high energy saving potential. Although it is a cost-effective greenhouse gas mitigation solution, budgetary constraints of its owners, who are often municipalities, lead to low refurbishment rates. To overcome this challenge, innovative financing solutions should be developed to attract other investors and reduce high up-front investment costs for the public budget.

The paper presents selected results of research, which aims to identify and suggest suitable models to upgrade public buildings and street lighting in Germany. It reviews financing models, which exist in Germany and neighbouring countries, including self-financing, debt-financing, third party financing, and public–private partnerships. The chapter further analyses these models using a common framework. In particular, it provides the overview of each model, identifies the projects to which it can be applied, specifies its advantages and disadvantages, as well as jurisdictions that have applied the model. The paper concludes with recommendations for German decision-makers on finding and implementing a suitable financing model.

A. Novikova (✉) · K. Stelmakh · A. Klinge · I. Juergens
Energy Efficiency and Climate Finance Department, Institute for Climate Protection, Energy and Mobility (IKEM), Berlin, Germany
e-mail: aleksandra.novikova@ikem.de

M. Hessling
European Lighting Expert Association (ELEA), Olten, Switzerland

P. Bertoldi (ed.), *Improving Energy Efficiency in Commercial Buildings and Smart Communities*, Springer Proceedings in Energy,
https://doi.org/10.1007/978-3-030-31459-0_11

1 Introduction

Investment in the upgrade of public infrastructure such as public buildings and street lighting offers significant energy savings and carbon dioxide (CO_2) emission reduction potential. Furthermore, it is often a cost-effective mitigation option with a short payback period. This raises the question why a significant share of this infrastructure in Germany requires refurbishment. Most often the problem lays in budgetary constraints of municipalities, who are most often the infrastructure owners. To overcome this burden, innovative business models are needed to raise additional, preferably private investment.

The paper aims to suggest suitable financing models to finance energy efficiency upgrades of public infrastructure in Germany with a special focus on public buildings and street lighting. For this, it identifies, reviews, and analyses financing models, which exist in Germany and neighbouring countries, including self-financing, debt-financing, third party financing, and public–private partnerships. Among these, the paper in particular discusses the models, which minimize the burden on the public budget. The paper builds on the results of Working Package 4.7 "Technical-Systemic Analysis with a Focus on Energy Efficiency in Buildings" of the System Integration Project—Enavi of the German Federal Ministry of Education and Research (BMBF).

After the introduction, the paper describes the methodology for the collection of financing models and the common framework against which they were analysed. Then, the paper provides an overview of all models and then for each model, it identifies the projects to which it could be applied, specifies its advantages and disadvantages, as well as jurisdictions that have applied the model. The paper concludes with recommendations for decision-makers on identifying and implementing a suitable financing model.

2 Methodology

There have been several data collection approaches, which we relied on. We interviewed the organizations that own, operate, and take decisions on the modernisation of public infrastructure in Germany and neighbouring countries of Central Europe such as Austria, the Netherlands, Italy, the Czech Republic, and France. These include municipal government structures, companies owned by municipality (e.g. utilities) and private or partially private companies delivering these functions. We also interviewed the organisations in these countries, which were involved in the financing of the modernisation of public infrastructure. These were energy service companies, manufactures of advanced technological components, public and commercial banks and institutional investors.

Further, we gathered information available in the public domain, for example from publication databases and project websites. Above all, we identified the projects, which have already conducted similar studies in the past. From our review, we concluded that so far there has been no recent comprehensive catalogue of the business models for energy efficient public infrastructure in Germany. Therefore, we identified individual case studies through Internet search and personal interviews. However, when a useful model was available only outside Germany, we included it, with priority given to models in the neighbouring countries, then the rest of the EU, and finally worldwide.

Then, each business model was analysed in a common framework. First, we provided a model overview based on its schematic structure (e.g. its key actors and their roles). Second, we analysed project types that could be financed using these models. Third, we argued about advantages and disadvantages of the models. Finally, for each business model, we selected a case study, in the context of which we developed greater details on the model, scope, stakeholders, implementation experience, and outcomes. In our analysis, we relied on the definitions developed by the OECD and Climate Policy Initiative for the analysis of energy and climate finance [1, 2].

After the information gathered was analysed, the draft report was distributed for comments among the project partners and the key stakeholders interviewed, when they had a chance to recommend outstanding business models and case studies.

3 Financing Models for Energy-Efficient Public Infrastructure

3.1 Review of Models Identified

As a research result, we identified a range of models to organize finance for the upgrade of energy efficiency in public buildings and street lighting. These include different alternatives of self-financing; debt-financing; third-party financing by a private contractor, a private contractor through energy savings, a utility, and citizens; and financing using public–private partnerships presented in Fig. 1.

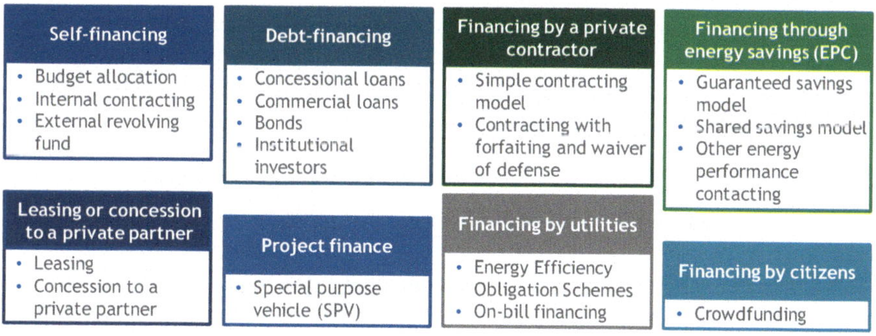

Fig. 1 Financing models for public infrastructure investment

3.2 Self-Financing

The most straightforward financing option for a municipality to upgrade the public infrastructure is to pay for it from own funds. In Germany and many other neighbouring countries, it is very common to use municipal budgets and dedicated national or federal funding sources to fund municipal infrastructure projects. The financing model of such upgrade project is rather simple (e.g. a municipality identifies a need for investment, prepares a funding request, obtains approval, and issues a tender, and selects a contractor to carry out the upgrade).

To minimize the burden on taxpayers, the public sector can design and implement additional schemes which help raise the funds for the budget. For instance, a municipality could invest in its infrastructure upgrades and once it accrues saved energy costs, it could re-invest them in new upgrade projects. These models are further discussed in the section.

Establishing an Internal Performance Contracting ("Intracting") Scheme

Model Overview

Internal performance contracting is a financing scheme in which organisational units within a municipality are contracted to implement a project without the need for external financiers. Energy efficiency measures are financed through reduced energy costs. The municipal budget supplies most of the initial financing, for example by setting up a dedicated revolving fund or trust. The fund or trust will finance energy efficiency or other emission reduction measures at 0% interest rate and without any extra charges. Creating a fund or trust of this kind requires political support and commitment from the department(s) responsible for the budget [3].

The "intractor" is another administrative unit (e.g. environmental agency, infrastructure department, or municipally owned company) that takes on the responsibility conventionally allocated to an energy service company (ESCO). The intractor assesses energy savings potentials, calculates the investment costs and payback period, and plans the project. It is crucial for the intractor department to have the skills and expertise required to prepare and implement such projects successfully. Once the project is implemented, cost savings gradually replenish the fund or trust until the investment has been paid off. The repaid resources are then used to finance new energy efficiency projects [3, 4] (Fig. 2).

Projects That Can Be Financed with This Model

This financing scheme can be used for municipal infrastructure projects, such as improvements in energy efficient street lighting or combined heat and power plants. The maximum project size depends on the total funding available from the finance department [4, 5].

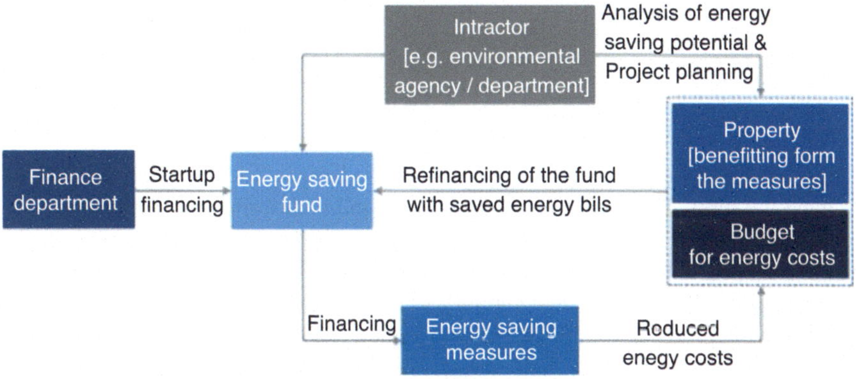

Fig. 2 Intracting model

Advantages

Intracting enables municipalities to use energy savings to finance energy efficiency measures without the involvement of external financiers. Cooperation between the units within a municipality thus helps to bridge the traditional divide between municipal investment and operational budgets. Projects that are too small or fail to attract private investment can be financed under this model with a 0% interest rate on investment capital. There are also minimal transaction and administrative costs, and the legal paperwork required is much simpler than that for external contracting [3].

Disadvantages

This model has the limitations generally associated with self-financing. For example, the municipality must provide the upfront capital. Municipalities use various approaches to sustain the fund and increase the finances available. They include the projects on the municipal balance sheet and bear all investment risks. Finally, projects financed by the municipalities may be implemented less efficiently than those funded by private investors [4, 6].

Jurisdictions That Have Applied the Model

Intracting was conceptualised and adapted in Germany, namely, by the city of Lörrach and Kiel. Cities in other European countries such as Agueda and Almada in Portugal, Udine in Italy and Koprivnica in Croatia are currently testing intracting in their jurisdictions [4, 7].

Establishing an Internal Revolving Fund and Outsourcing Service Provision

Model Overview

Under this model, a municipality or municipal bodies provide the initial capital and manage the fund. The revolving nature of the fund is described in the introduction to this section. The revolving fund provides finance (grants, loans, or other financial instruments) to external service providers and ESCOs to implement energy efficiency projects. The replenished capital and energy savings are reinvested into new energy-saving projects. If municipalities are small and do not have enough resources to set up an individual own fund, they could combine their resources to create a common revolving fund with shared management and operating costs.

Projects That Can Be Financed with This Model

The model could be applied within medium-sized to large cities or groups of smaller municipalities that need to finance long-term and multi-objective projects, including those for building energy efficiency improvements, street lighting, combined heat and power plants, and renewable energy. The maximum project size will depend on the size of the total budget available from the fund. However, cumulative energy savings should be significant enough to justify the complex set-up required for the fund and its operating costs.

Advantages

The key advantage of the model is its sustainability and long-term orientation. Once a revolving fund has been established and energy savings accrued from the first investment projects, the resulting energy savings can be used for new projects. Ideally, interest rates, fees, service charges, and/or energy cost savings could completely cover fund operating costs [8, 9]. An internal revolving fund provides an alternative financing mechanism for municipalities that have limited borrowing capacity [10].

Disadvantages

Establishing a revolving fund requires significant political commitment, as well a sizeable investment of time and institutional and human capacity. Cash inflows will accrue only after several years, so recovering fund operating costs may be a lengthy process [8]. Therefore, financial sustainability is important to ensure the continued cost-effectiveness of the fund, as well as the long-term strategy for resource mobilisation. Furthermore, a revolving fund requires an entity to manage the funds to ensure good governance and administration. Finding a dedicated and experienced staff to support and operate the fund is sometimes difficult, especially in small municipalities [9].

Jurisdictions That Have Applied the Model

The energy saving fund implemented in the Czech town of Litomerice is one example of an internal municipal revolving fund [11].

Establishing an External Revolving Fund with Multiple Financiers and Service Providers

Model Overview

Alternatively, a revolving fund can use external funding sources and provide finances to municipalities for energy efficiency projects. The money to operate the fund and supply the first and future investments could originate from one source or a combination of sources. Potential sources include grants and/or loans from public and private sources, such as the national or regional government, financial institutions, utilities, energy service companies, and/or other capital providers. If a fund of this kind becomes self-sustaining over time, it can lend repaid capital to new projects and finance its operating costs from service charges and interest on loans [8]. The external fund is often managed by a dedicated fund manager, which could be a specially created new entity, a utility, an ESCO, or another organisation [9]. The municipality contracts a third party to carry out an infrastructure project, and repays loans from saved energy costs. The contracts may be conditional on the contractors' compliance with performance standards.

Projects That Can Be Financed with This Model

As is the case for internal energy efficiency funds, the success of this financing model depends on the available funding and programmatic priorities of the specific municipality.

Advantages

One obvious advantage of external revolving funds is that, by combining sources (financial institutions and private investors in particular), they make it possible to obtain greater funding. Private investors can participate in urban development projects that showcase energy efficiency benefits to citizens and communities [9].

Disadvantages

The process of setting up a revolving fund, particularly an external one, requires the cooperation of various stakeholders and involves several steps, which increases the complexity of the financing model. The role of a private entity as a fund manager of a partially public fund might pose a political challenge, especially when such fund has the potential to act monopolistically [9]. If a revolving fund uses private and public capital, tensions may arise as a result of the transfer of most responsibilities from donors to fund managers (e.g. project selection) [12].

Jurisdictions That Have Applied the Model

An example of external revolving funds at the municipal level is an Energy Fund Hague [13].

3.3 Debt-Financing

Many municipalities with limited own funds issue debt that it later pays back from the municipal budget, for example from municipal tax revenue and/or saved energy costs. Under the debt financing model, the municipality incurs debt (e.g. by taking out loans or issuing bonds) and then issues a tender for a contractor to perform the upgrade.

Concessional or Commercial Loans from Banks

Model Overview

Most municipalities obtain low-interest or concessional loans. European municipalities often finance energy efficiency investments with low-interest-rate lending programmes through national development banks, dedicated funds, or European banks and funds (e.g. the European Investment Bank—EIB) in cooperation with local commercial banks. In the countries where such programmes have limited budgets or do not exist, municipalities obtain market-rate loans from commercial banks. In such cases, the interest rate does not depend on saved energy costs but on the borrower's credit record.

Advantages

Concessional loans allow municipalities to access capital at below-market interest rates and, unlike commercial loans, do not require proof of a flawless credit record. Capital from debt financing can be combined with money obtained from other financing models, including revolving funds, with minimal administrative requirements.

Disadvantages

Although interest rates are typically low, the municipality must repay the debt. In addition, because the debt is listed on the municipal balance sheet, it weakens the equity-to-assets ratio, with negative effects on the municipality's access to credit.

Projects That Can Be Financed with This Model

If a municipality has a positive credit record, any of its projects can be financed under this model.

Jurisdictions That Applied the Model

In Germany, a dedicated programme of the Kreditanstalt für Wiederaufbau [Reconstruction Credit Institute—KfW], a national development bank supported by the German government, offers loans for developing sustainable urban infrastructure and public buildings at interest rates close to 0%. It is the main source of funding available to German municipalities. Multiple loan programmes and other financial instruments are available through European institutions and intermediaries, such as the EIB. EU member states are also encouraged to use grants from European Structural and Investment Funds (ESIF) to create other financial tools (e.g. loans, guarantees, equity, or risk-bearing instruments) to improve the efficiency of EU resources.

Issuing Municipal Bonds

Model Overview

Municipal bonds are issued by a local government or their agencies to raise funds for investment projects. When a municipality issues a bond, it acquires a debt obligation and must pay interest and/or repay the debt in the future. While municipal bonds can be used to finance any type of municipal investment, labelled green bonds are issued exclusively for projects related to sustainability. To have a "green" label, bonds must be certified by an independent institution.

Projects That Can Be Financed with This Model

Most municipalities with access to a bond agency can apply this model.

Advantages

The municipality can raise finances for public projects independently or in coordination with a bond agency. Because bonds generally have low interest rates, it offers capital at a lower cost than do commercial loans.

Disadvantages

Issuing municipal bonds may require extensive and costly preparation, such as obtaining a credit rating, acquiring approval from national securities authorities, and consulting with investment brokers. For this reason, many countries have municipal bond agencies, which aggregate debt from multiple municipalities, issue bonds, and sell them in the financial markets. With a high credit rating, agencies can raise capital for municipalities at a lower cost than would be possible if the municipalities issued the bonds themselves. Agencies of this kind are located in Sweden, Finland, France, Denmark, Switzerland, UK, and the Netherlands [14].

Jurisdictions That Have Applied the Model

Bonds are less common in Europe, but widely used in the USA. The European examples include Gothenburg, the second largest city in Sweden. Gothenburg has started the Gothenburg Green Bond Program, which provides funding for sustainable investments in the city infrastructure. The American examples include for instance federal and state level programs Qualified Energy Conservation Bonds (QECBs) as well as initiatives by the individual cities. Some examples including the cities of Detroit and San Diego, where bonds were used to raise finance for modernisation of public infrastructure [15, 16].

3.4 *Financing by a Private Contractor*

Alternatively, municipal actors can transfer the burden of financing energy efficient public infrastructure to third parties, for example by contracting an ESCO. The section focusses on models in which financing is provided by a private partner (e.g. an ESCO) that performs the upgrade but is generally not responsible for the energy supply, which prevents it from using energy savings to recoup its own costs. The private partner finances the energy efficiency project from its own funds or obtains third-party funding. Municipalities pay the contractor for its services. If the contractor also obtains financing from third parties, it repays its debt to them. Neither the municipal payment to the contractor nor the contractor's payments to the lender are contingent on energy savings.

Simple Contracting Model

Model Overview

In a simple contracting model (see Fig. 3), the contractor may have several responsibilities, but its main activities usually include the planning, financing, and execution of the investment in new, energy-efficient public infrastructure. The contractor may also be responsible for operating this infrastructure, although this is generally not the case for several reasons. First, the municipality either uses its own resources (potentially including subcontractors), or has contracted an external operator to perform this task, often over a longer period. Such contracts often do not cover the use of new technologies for retrofits and modernisation. As a result, the municipality must wait until the end of the contract period to issue a new tender including modernisation, or it must find a different partner to carry out the task. Because the time most suitable for the investment rarely coincides with the end of an existing service contract, finding a different partner for the modernisation is often the most practical solution.

In a simple contracting model, the contractor receives a contracting fee directly. This fee covers the costs of planning, financing, and executing the investment. The length of such contracts may vary in accordance with various factors, including the size of the investment and its relation to energy costs. The contract often must be put out to tender, and there are various evaluation procedures for offers received from vendors. For example, the municipality may define framework conditions, like the minimum energy savings to be achieved, details about technologies to be used, terms of warranty, and performance standards. In addition, the procedure for the end of the contract must be defined. For instance, for street lighting, the contractor should be able to remove the installed technology when the contract expires. Due to the long duration of such contracts, however, it is very unlikely that this will take place, because contractors will not be able to sell relatively old components for further use.

Tenders are generally evaluated on the basis of the municipal potential savings. Savings can be significant, particularly in countries with relatively high energy prices. In such cases, the contracting fee, which the municipality pays to the contractor, is often much lower than the energy savings achieved. Maintenance costs are often reduced as well, because little upkeep is required for new technological components.

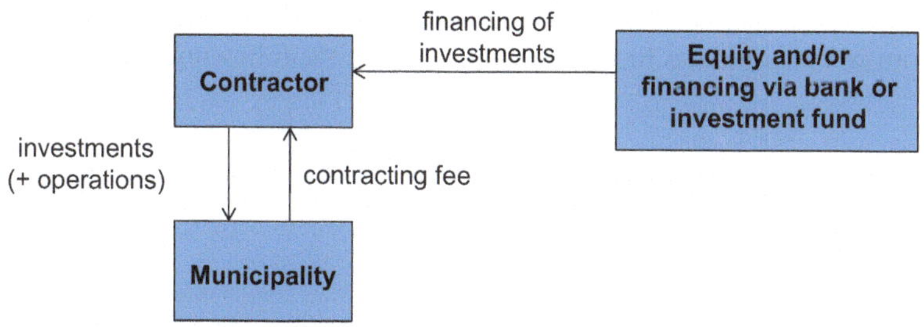

Fig. 3 Simple contracting model

Projects That Can Be Financed with This Model

Projects must meet minimum size criteria to justify the contractor's participation in the project, which often requires co-financing by a bank. There is no fixed size threshold, but a project volume of EUR 0.5–1 million is a reasonable minimum. The value must be raised, if the contractor also becomes responsible for carrying out operations, because in this case it must establish an office with personnel and equipment in or near the municipality.

Advantages

The key advantage is that the model is off-balance sheet for the municipality. Contracting models differ from loans in this respect because the latter are normally on-balance sheet for the municipality. The responsible authority for the municipality should be asked to verify use of the off-balance model, however. A further advantage is that the tendering process provides an opportunity to select specialised companies that have relevant know-how and experience and offer more attractive prices than those of the municipality or present operator.

Disadvantages

From the municipality's perspective, the major disadvantage of the model is its high financing costs. The capital costs of contracting—whether through equity or re-financing loans from banks—are usually significantly higher than those payable through direct financing from the municipal budget or financing models with low interest loans. From a contractor's perspective, the off-balance sheet financing is a disadvantage, although private investors generally favour off-balance sheet models.

Jurisdictions That Have Applied the Model

Simple contracting models are widely applied for public infrastructure projects, although both partners typically search for ways to reduce financing costs by applying more complex models like the one described in the following section.

Contracting Model with Forfaiting and Waiver of Defence

Model Overview

In this more complex model, the municipality and contractor play roles similar to those in the simple contracting model described above. In contrast to that model, here bank involvement is a central element and the bank enters into agreements with

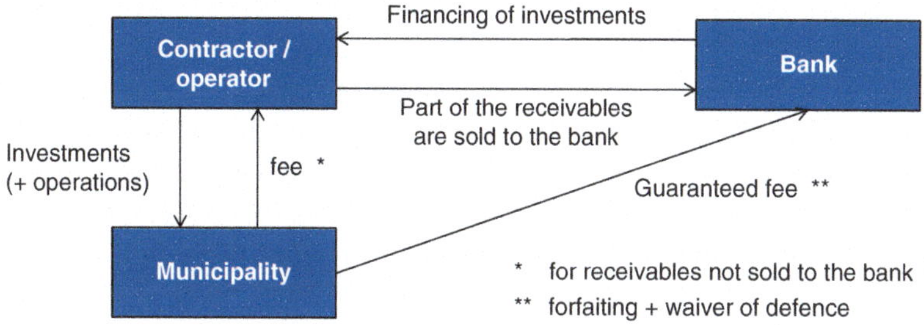

Fig. 4 Contracting model with forfaiting and waiver of defence

the contractor and the municipality (Fig. 4). The contractor sells part of its future receivables to a bank in a "forfaiting transaction". The municipality must then pay a portion of the contracting fee to the contractor and another (typically larger) portion directly to the bank, or "forfaiter". The portion of receivables sold to the bank corresponds to the value of the equipment installed and is therefore higher than the portion for the planning, installation, and warranty.

The bank participates in the risks arising from uncertainties associated with the receivables and earns a margin. In this particular model, the municipality may provide a guarantee to the bank, referred to as a "waiver of defence". This guaranteed fee minimises the bank's risks, even for the worst scenario (e.g. no energy is savings achieved). The contractor bears all other risks, including those associated with the equipment. The contracting model with forfaiting and waiver of defence is therefore a mechanism to generate funds while minimising risks to the bank, thereby reducing the interest rate to a level typically available only to municipalities. The rate reduction may seem modest but can add up to a substantial amount over the long duration of the contract.

As in the previous model, the contractor owns the installed energy efficient equipment for the contract term. When the contract expires, there should be an option for the contractor to remove the components. If no such option exists, a leasing contract should be created to regulate any further municipal use.

Projects That Can Be Financed with This Model

There is no big difference to the simple contracting model, as far as minimum project size is concerned. Due to the higher complexity of the model, it might be a bit more difficult to find a bank financing projects below EUR 1 million.

Advantages

This model has similar advantages to the simple contracting model. Furthermore, contracting with forfaiting and waiver of defence will have a lower interest rate, included in the contracting fee.

Disadvantages

Although the interest rate is lower than in a simple contracting model, it still is much higher than in the case of financing through models with low interest rate lending programmes, such as the KfW loans currently available in Germany at close to 0% interest rate. A further disadvantage is the higher complexity of this model, as well as the fact that a large part of the municipal payments, namely the payments to the bank, have to be guaranteed, regardless of the performance.

Jurisdictions That Applied the Model

There is limited experience with the application of this model so far. We located only two case studies using the model. These are the city of Dillinburg, Germany and the city of Lytomysl, the Czech Republic.

3.5 *Financing by Private Partner Through Energy Savings*

Energy performance contracts (EPCs) are used to finance municipal infrastructure projects by a private partner, usually an ESCO, through energy savings. There are various EPC models, but the element common to all types is that the costs saved through reduced energy consumption are used to finance the investment. In some EPC models, the municipality may be responsible for the energy supply, while in others, the private partner may be the responsible party. Ultimately, however, the municipality always pays, in a per-service fee or a lump sum, for the operations, including the energy supply, planning, financing, and installation of the new equipment. EPC contracts are especially common in Germany, Denmark, and the UK; they are also often used in Finland, the Czech Republic, and the Netherlands [17].

Energy Performance Contract: Guaranteed Savings Model

Model Overview

In an EPC with guaranteed energy savings, the ESCO designs and implements the project and is obliged to achieve a certain level of energy savings. If the ESCO fails to deliver these savings, it must cover the shortfall. If the ESCO delivers higher energy savings than guaranteed under the terms of the contract, these savings fully benefit the municipality. Over the contract term, the municipality pays a fixed fee from the cost savings.

Energy savings should, however, be sufficient to pay for the modernisation in a reasonable time. This already shows a potential problem of this model. In countries with low energy prices, the payback period may be so long that it is difficult to attract private partners willing to enter into such contracts. This was the case in some Eastern European countries, where energy prices can fall below 10 EURct/kW h. As a result, the duration of EPCs could exceed 20 years, while in a typical Western European country with energy prices close to or above 20 EURct/kW h, much shorter contract lengths are possible.

In the model shown in Fig. 5, future municipal costs (consisting of energy costs plus regular payments to the private partner) are identical to municipal energy costs paid before the modernisation took place. Sometimes energy efficiency upgrades can deliver energy savings of up to 80% or more, for instance in case of the retrofit of street lighting infrastructure. In this case, municipalities can use a significant portion of the cost savings to cover EPC fees to the service provider. This arrangement can either shorten the length of the contract or reduce the municipality's regular payments, allowing for immediate savings even during the contract term (Fig. 6).

In all EPC models, the private partner guarantees a certain level of energy savings. This helps the municipality by enabling it to calculate future costs with a high reliability, because the private partner bears the risks associated with achieving the energy savings. There is generally also an agreement to reduce payments to the private partner if it fails to provide the guaranteed savings. While this theoretically constitutes a risk to the private partner, the guaranteed energy savings proposed in its initial offer may include a buffer to absorb such shortfalls.

Projects That Can Be Financed with This Model

This model is suitable for projects with the potential to deliver significant energy and cost savings. Otherwise, the contract duration may be too long to attract private partners. In addition, municipalities should have sufficient financial resources to pay the same amount (or slightly less) over the length of the contract, even though payments are split to cover energy costs and payments to the private partner.

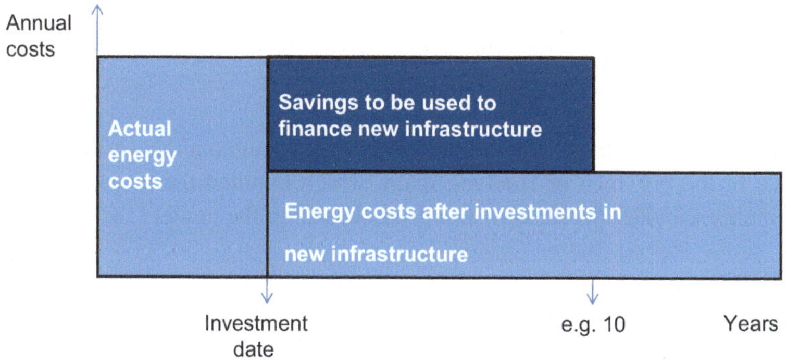

Fig. 5 EPC—guaranteed savings model (time-optimised)

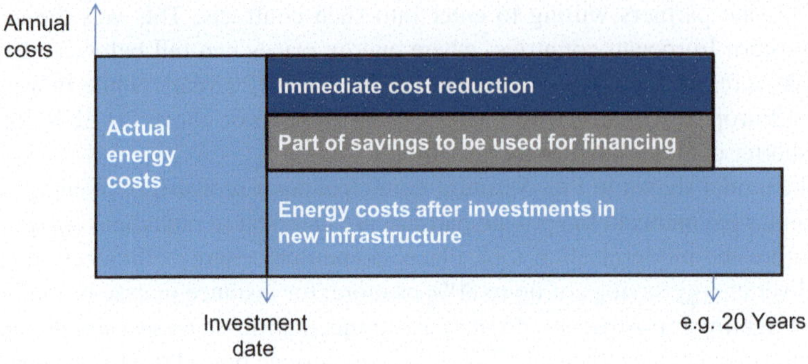

Fig. 6 EPC—guaranteed savings model (with immediate cost reductions)

Advantages

The primary advantage is that the municipality will complete energy efficiency upgrades without sharp upticks in public spending. The amount paid by the municipality over the contract period is equal to or less than the amount that it paid before the EPC; after the contract expires, the municipality owns energy-efficient infrastructure and benefits from the lower operating costs. A further advantage for the municipality is the fact that the major risks are borne by the private partner.

Disadvantages

Some disadvantages are indicated above, including those associated with low energy prices and/or the efficiency of the existing infrastructure. Another disadvantage of this model is that it does not incentivise the private partner to reduce energy demand to a level below that guaranteed in the contract. This problem can be solved with the "shared energy savings" model discussed below.

Jurisdictions That Have Applied the Model

Several variations of this model have been applied throughout the whole Europe, for instance in the province of Huelva, Spain, which bundled the projects of several municipalities and tendered them as a group based on the model [18].

EPC: Shared Savings Model

Model Overview

In this model, the energy savings obtained are greater than those guaranteed under the terms of the contract between a municipality and private partner, and the parties share the additional savings. This is a "win-win" situation when the contractor and municipality are interested in generating the greatest possible energy savings. It is thus more common than the previous model.

The contract generally specifies a level of guaranteed energy savings, as well as a malus agreement that cuts payments to the private partner if the guaranteed savings are not delivered. In addition, the municipality and the private partner share any savings in excess of the amount guaranteed. The bonus payment to the private partner can be a certain price (in EUR/MW h) or a portion of the saved energy costs, based on an electricity price agreed upon by both partners. This model provides a clear incentive for additional energy savings on both sides, with the savings shared as prescribed under the agreement (Fig. 7). The parties may split the savings 50/50, although this is not always the case. In the tendering process, bidders can be instructed to propose a distribution of any extra savings.

Projects That Can Be Financed with This Model

The criteria for EPC guaranteed savings models also apply here. The shared savings model is appropriate only if potential energy savings are high; otherwise, the amortisation period is too long.

Advantages

One major advantage of this model is that both sides have an incentive to consider and realise additional energy savings, even if these were neither planned nor anticipated. Municipalities can invest their share of additional energy cost savings into energy efficiency projects. The model also offers the same advantages as the guaranteed savings model.

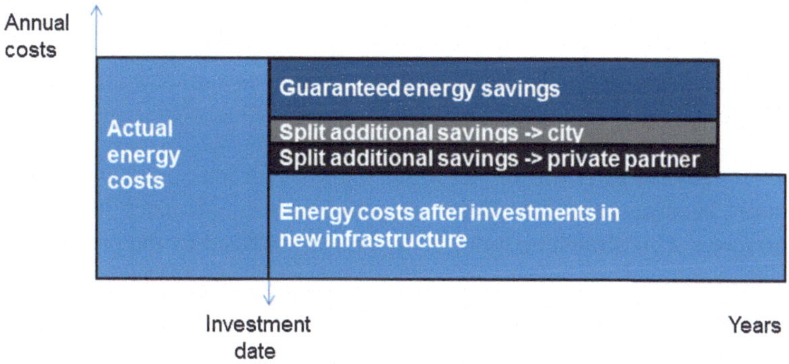

Fig. 7 Energy performance contracting—shared saving model

Disadvantages

One problem with EPC guaranteed savings models in which savings are not shared is that they do not incentivise contract parties to realise energy savings above guaranteed levels. Although the shared savings model resolves this problem, it does not mitigate other disadvantages, including the long payback periods that result from low energy prices.

Jurisdictions That Have Applied the Model

Similar, several variations of this model have been applied throughout the whole Europe, for instance in the city of Nauen, Germany.

EPC: Modernisation with Immediate Savings of Energy Costs

Model Overview

Sometimes improvements in the infrastructure can be completed over a few months, allowing for immediate savings in energy costs. This EPC model can provide for guaranteed savings only or shared savings.

Projects That Can Be Financed with This Model

Models that call for immediate modernisation (for instance, the replacement of old luminaires with new in the shortest time) are appropriate if installed technologies are old and no longer acceptable in terms of their efficiency, reliability, and maintenance costs.

Advantages

The primary advantage of this model is that it allows for maximum energy savings. In addition, because new technologies—require less maintenance, the associated costs are lower. This should be reflected in the price offered by the private partner.

Disadvantages

If the municipality covers all modernisation costs, the sizable initial investment may be prohibitive. The terms of the agreement also require the municipality to tolerate extensive construction activity at one time. The municipality must consider the consequences (for example, with regard to traffic congestion and public reaction).

In long-term contracts, the early realisation of savings means that, by the time the contracted work is complete, the entire infrastructure is once again outdated and must be replaced. The modernisation process completed over the contract period will not incorporate any new technology; this prevents the municipality from modernising infrastructure at a constant rate (for instance for street lighting, cities typically modernise 3% of existing infrastructure per year with the most advanced technology). The staggered modernisation model eliminates these disadvantages by providing for time-optimised (i.e. longer-term) modernisation and utilisation of energy savings.

Jurisdictions That Have Applied the Model

The EPC guaranteed savings model with immediate energy cost savings has been implemented in several European cities, for instance, in the city of Tona, Spain [19] and by the Bulgarian municipality of Sapareva Banya [20]. Berliner Energieagentur GmbH reported the successful application of this model to complete ten projects with the Clearinghouse for Energy Contracting in Central and Eastern Europe in 2003–2004, as well as 17 such projects in 2005–2007 with EuroContract Guaranteed Energy Performance [21].

EPC: Model with Staggered Modernisation

Model Overview

The staggered modernisation model (see Fig. 8) addresses the disadvantages of the previous model by extending the modernisation process over a longer time period. Over several years or even decades, outdated technologies are retrofitted with energy-efficient gradually. At regular intervals during the contract term and again at the end of contract, the municipality and private partner can establish a modernisation time schedule or specify the maximum age.

Projects That Can Be Financed with This Model

Models with staggered modernisation are appropriate in cases in which the performance varies among the existing technology. It makes sense to replace the oldest and least efficient components first, and wait with others until their technical and economic performance indicates that they have reached the end of their useful life.

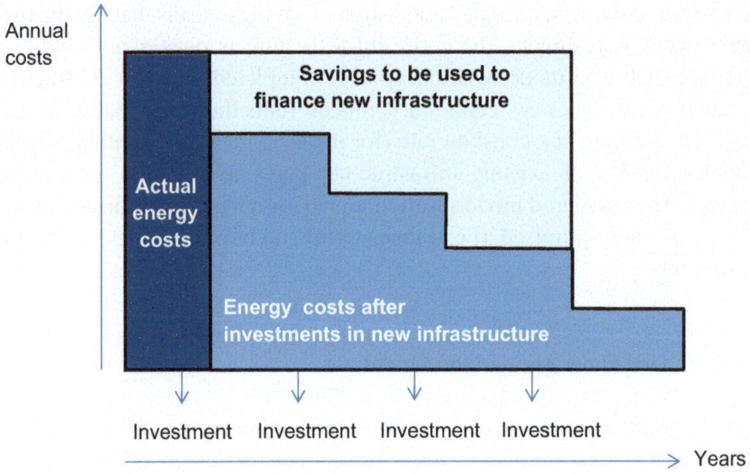

Fig. 8 EPC—Model with staggered modernisation

Advantages

The advantage of this model is that it provides for regular investments. As a result, the municipality maintains reasonably modern infrastructure, and there are no sharp upticks in investment requirements or building activity. The model also helps to prevent a situation in which all components are replaced at the same time but the least energy-efficient can be replaced first.

Disadvantages

The primary disadvantage is that the municipality realises the benefits of energy savings and lower maintenance costs later than it would if it applied the immediate savings model.

Jurisdictions That Have Applied the Model

This model has been implemented successfully in Germany for instance for the upgrade of street lighting of the city of Hilden.

EPC: Model with Related Payments

Model Overview

Under an energy performance related payment (EPRP) contract, an ESCO guarantees to improve the energy efficiency of public infrastructure with guaranteed savings for an agreed fee. The fee is contingent on energy savings, which means that if savings fall short, the ESCO loses a portion of its fee [22].

An EPRP contract specifies guaranteed energy savings, sets a fee, and establishes a payment mechanism for measured and verified savings (see Fig. 9). The EPRP model can be added as a clause to traditional contracts or tender documents as a guarantee that the project will fulfil its intended objective in a way that improves energy efficiency. The performance risk is shared between the municipality and the ESCO [22]. The contract generally expires 12 months after the work is complete. After this point, the vendor is unlikely to finance the infrastructure.

Projects That Can Be Financed with This Model

This model, like the previous EPC models, is best suited to projects with the potential to generate significant energy savings such as street lighting.

Advantages

Payments based on energy efficiency lead to a more accurate quantification and verification of energy savings. ESCO's objectives align with those of the municipality objectives, because the ESCO focusses on the energy performance of the upgrade

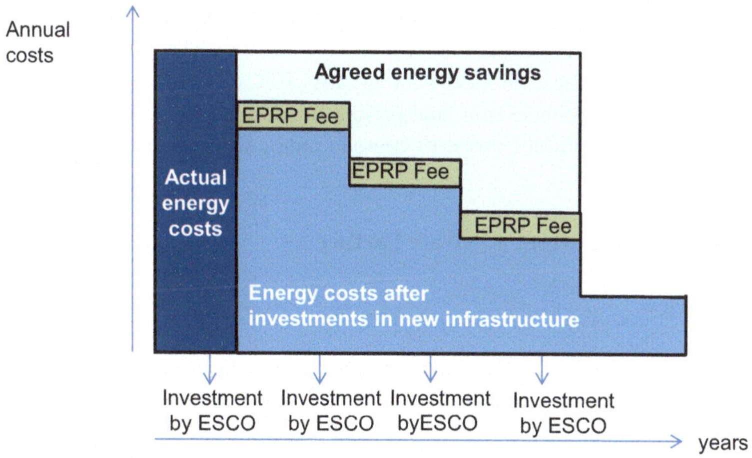

Fig. 9 EPC—Model with energy performance related payments

after the final payment. Because of the related payments to energy savings, funding is more easily accessible [22].

As in other models involving private actors, the technical expertise of the ESCO is likely to surpass that of the municipality and is therefore able to perform upgrades that guarantee energy savings. Another advantage is the nature of the contract: in its simplest form, the EPRP method is similar to traditional contracting [22] and can be combined with other funding models, including those based on grant or debt financing.

Disadvantages

Like the staggered modernisation model, the EPRP prevents the municipality from realising early cost savings, because some of the upgrades are performed at a later stage.

Jurisdictions That Have Applied the Model

Multiple case studies of the EPRP model are located in various parts of Ireland. For example, the Port of Cork and the Dublin Port headquarters have applied the model to lighting infrastructure [22].

3.6 Public-Private Partnerships

In recent years, public and private actors have created new types of infrastructure contracts. These actors enter into contracts to share responsibilities associated with infrastructure project implementation and/or operations management. These models are referred to as "public–private partnerships" (PPPs). The technical and financial specifications of these models vary greatly. EPCs, discussed above, are one form of PPP. Leasing, concession, and private finance models, outlined below, are other PPPs used throughout Europe to upgrade public infrastructure.

Sale- and Lease-Back from a Private Partner

Model Overview

A party leases an asset when it grants another party (the "lessee") the temporary right to use an asset without purchasing it. Sale-and-leaseback models are often used to finance the energy efficiency upgrade, operation, and management for a certain amount of time. A municipality sells public infrastructure to a private contractor conditional on upgrade, operation, and management. The municipality

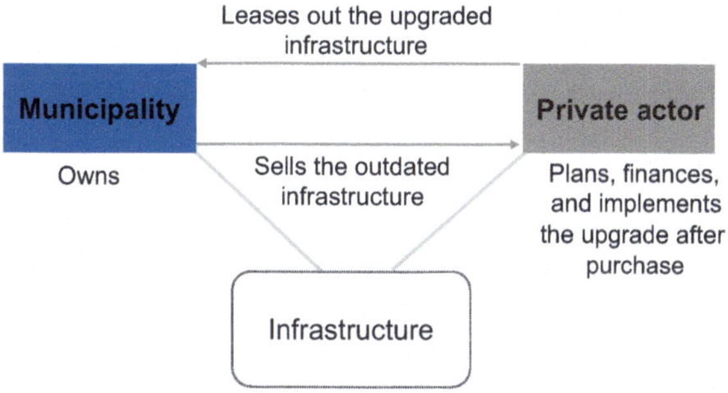

Fig. 10 Leasing model for street lighting between a private partner and a municipality

than leases it back from a private contractor for a fixed fee over a set period of time. Ownership rights are generally transferred back to the municipality at the end of the leasing contract. A sample leasing model is presented in Fig. 10.

Projects That Can Be Financed with This Model

Leasing is appropriate when upfront investment costs for a project are too high for the municipality to self-finance. When property is leased, the owner bears the investment costs; the municipality reaps the benefits of upgraded infrastructure without increasing its debt. Depending on the specific provisions of the leasing contract, the municipality may be permitted to repay the lease with the savings generated [23].

Advantages

The financial risks and costs are spread out over time. Partners to a leasing contract share the risks related to asset performance and maintenance.

Disadvantages

The municipality has less control over the assets. In the long term, leasing is costlier than self-financing.

Jurisdictions That Have Applied the Model

The leasing model is not common in Europe but has been implemented for instance in the Italian municipalities of Cesena and Martignacco. It has been applied by multiple jurisdictions in the USA, including in Pennsylvania and Texas [23]. Also the city of Guadalajara, Mexico used the leasing model, combined with governmental support, to upgrade public infrastructure [24].

Concession to a Private Partner

Model Overview

Under a concession model, the municipality commissions a private partner ("concessionaire") to operate and maintain infrastructure over a set period. The municipality pays the private operator a fee for this service. To cover costs, the private partner can use its own funds or raise debt in the capital market and invest the balance in energy-saving measures. By using these funds to reduce the costs of operation and maintenance, the contractor can accrue revenue. In response to Directive 2014/23/EU on the award of concession contracts, each EU member state implemented national legislation regulating tenders for concession contracts. Figure 11 presents a simplified model of a concession agreement between a municipality and a private partner.

Projects That Can Be Financed with This Model

Any municipality can outsource the infrastructure operation and maintenance to a private-sector company by drawing up a concession agreement in line with the legislation of the respective national government.

Advantages

Under a concession contract, a private company is responsible for operation and maintenance. It also must invest in upgrades and bear associated investment risks, including those related to technical aspects of implementation. Another potential advantage to the municipality is that it can stimulate innovation by setting high energy efficiency standards in the tender for contracts [25].

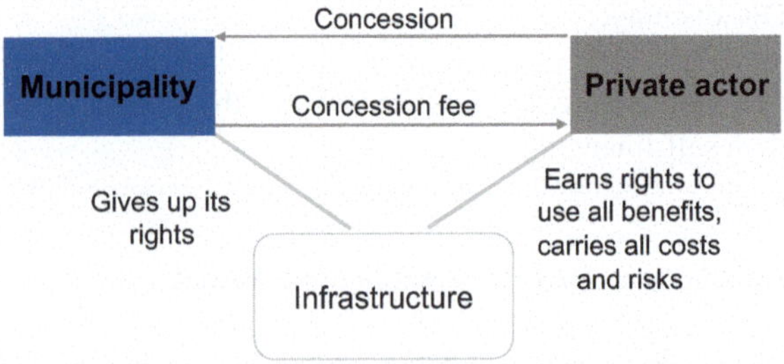

Fig. 11 Sample concession-based model between a private partner and a municipality

Disadvantages

Setting up and administering a concession model can be complicated. In particular, the negotiation and tendering phases may require significant manpower and time. Therefore, the transaction costs can be substantial. In addition, once the contract is signed, close regulatory oversight is required [25].

Jurisdictions That Have Applied the Model

In the EU, concession models have been implemented in the north eastern Italian municipality of Azzano Decimo [26] and in Paris, France [27].

Project Finance

Model Overview

The project finance model is another method by which municipalities can leverage limited public funds and raise private capital for PPP projects. The project participant creates a special purpose vehicle (SPV), a subsidiary legal entity that facilitates financial objectives while minimising the parent company's risk exposure. Its balance sheet documents project expenditures (Fig. 12). The SPV is financed through equity from private investors, debt from lending institutions, and contributions from the municipality [28].

Project finance has been implemented successfully for large transportation and energy infrastructure projects [29]. There is growing interest in this model, and its use for urban-scale energy efficiency projects is becoming more common; however, to date, there is no standardised approach to its implementation [8].

SPV arrangements include one or several private sector partners, such as equity investors, manufacturers, debt providers, and asset operators. The SPV is responsible for designing, installing, operating and managing energy efficient infrastructure at its own expense for the life of the contract. During this period, partners from the private sector bear most risks associated with asset ownership [28].

Fig. 12 Example structure of SPV model

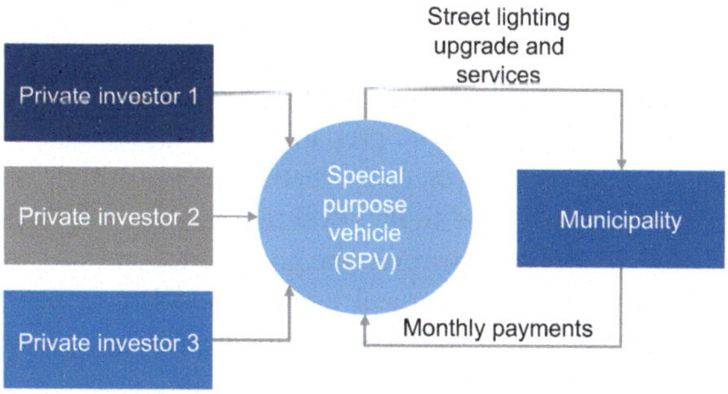

Contract terms are generally 20–25 years. The contract price is based on the required investment, cost of capital, and operation and maintenance costs. The municipality pays private-sector partners monthly unitary charges, the amount of which is based on the contract price. These payments represent the key security for funders [30, 31].

Projects That Can Be Financed with This Model

The project finance model is suitable for large projects with capital costs exceeding approximately EUR 20 million. The projects must be able to attract private investors to ensure financially sustainability. Financial sustainability depends on the revenues and profit realised over the contract term and is linked to the credit profile of the municipality. Supporting public financing tools like grants, tax exemptions, tax-free bonds, or credits can significantly improve project viability and facilitate private-sector involvement. The project finance model also allows for long-term contracts with private actors for the operation and maintenance [30]. The simple regulatory structure, clear legislative provisions, and fast, transparent bidding process are prerequisites for successful project implementation [28, 32, 33].

Advantages

The primary advantage of this model for municipalities is that it provides an opportunity for them to leverage private capital and implement projects using off-balance financing. The off-balance structure is also an important advantage for investors, manufacturers, and operators. From the perspective of both the public and private sector, one major advantage is that the SPV isolates project risks, enhancing the attractiveness of the investment. Long-term contracts provide stability for operations and asset maintenance [28, 34]. An additional benefit for municipalities is that, if private-sector partners fail to supply the services as agreed in the contract, the municipality is entitled to deduct or withhold a predetermined amount from payments and can even impose penalties.

Disadvantages

The main challenge in using the project finance model is absorbing the high transaction costs associated with preparing and implementing the SPV. This model is not suitable for small projects. Creating a consortium of several municipalities can increase project scale to the level required for a SPV; it also diversifies investment portfolio and risks. However, the governance and structure of the consortium involve additional costs. The project finance model may also allow a long time to elapse between the project start date and the beginning of actual development [28, 35].

Jurisdictions That Have Applied the Model

The project finance model has been widely used for energy efficiency investments across the UK in Private Finance Initiative (PFI) and Private Finance 2 (PF2) projects. As of March 2016, 32 UK jurisdictions had applied SPV models for public infrastructure investments with an average capital value of £45 million (EUR 57 million) [36]. The project finance model is also common in Italy, where it has been used by municipalities including Udine, Codroipo, Spilimbergo, Mereto di Tomba, and Morsano al Tagliamento [26].

3.7 Other Models

Financing by Citizens

Crowdfunding is a relatively new financing option and is most often used by young, innovative companies and start-ups for small or medium-scale projects. It involves raising funds from a large number of individuals or small-scale investors through online platforms. Crowdfunding creates a community around the project, allowing people to become more engaged and to provide useful insights and ideas to improve the project. Use of this instrument for community and city projects is also increasing [37]. The German crowdfunding platform Bettervest solicited funds of EUR 420,000 from 470 investors to retrofit the heating system of a school complex in Wilhelmstadt district of Berlin [38].

4 Conclusion

Even though the retrofit public buildings and street lighting offers high energy savings, its rate is low in Germany and in many other geographical jurisdictions in Europe. High up-front investment costs are among the highest barriers for municipalities. Therefore, the authors of the paper reviewed and analysed different financing options placing special focus on the third-party financing.

The most straightforward financing model is financing infrastructure upgrades from the own funds of municipalities. To minimize the burden on taxpayers, the public sector could design and implement additional schemes which help raise the funds to the budget, in particular a revolving scheme. Many municipalities, whose own funding resources are limited, obtain debt which is then be paid back from the tax revenue of municipalities and/or saved energy costs. Thus, the municipalities could obtain a low interest loan from a public lending program, a commercial loan from a commercial bank, or it could issue municipal bonds.

Municipal actors can reallocate the burden of financing the public infrastructure on third parties (e.g. contracting an energy service company). There is a wide variety of such contracts. In a simple contracting model, the contractor directly receives a contracting fee, which covers the costs of planning, financing and execution of the infrastructure upgrade, as well as its margin. The city is not involved into the contractual relationships with a bank providing the funds. In a more complex model with forfaiting and waiver of defence, the roles of the city and the contractor are similar to the simple contracting model, but the bank enters into agreements both with the contractor and with the city.

The other configuration is a set of energy performance contracting models, which could be applied when either a municipality or the contracted party pays for energy supply. In this model, the energy cost savings achieved via a reduction of energy consumption are used to finance the retrofit. Typically, the contracted energy service company guarantees an energy saving level to be achieved. In some models, additional energy savings achieved on top of the guaranteed level are shared between the municipality and the contractor.

In recent years, public and private actors have created new types of infrastructure contracts, public–private partnerships, sharing responsibilities associated with infrastructure project implementation and/or operations management. The technical and financial specifications of these models vary greatly. Energy performance contracts, discussed above, are one form of public–private partnerships. Leasing, concession, and private finance models are other public–private partnerships used throughout Europe to upgrade the public infrastructure.

Each of the models has its advantages and disadvantages as well as constraints to do with the economic, market, and legal conditions in which it could be applied. Therefore, the choice of model should be made according to the specific conditions in each municipality.

Acknowledgements Research conducted for this paper was a part of Working Package 4, Task 7 "Technical-systemic analysis with a focus on energy efficiency in buildings" of the System Integration project under the Kopernikus platform (ENavi—https://www.kopernikus-projekte.de/projekte/systemintegration). The project was financed by the German Federal Ministry of Education and Research (BMBF).

References

1. Buchner, B., Falconer, A., Hervé-Mignucci, M., Trabacchi, C., & Brinkman, M. (2011). *The landscape of climate finance*. Venice: Climate Policy Initiative.
2. OECD. (2016). *OECD glossary of statistical terms*. Paris: Author. Retrieved from https://stats.oecd.org/glossary/index.htm.
3. Energy Cities. (2016). *Internal contracting (intracting). The City of Stuttgart. Fact sheet*. Besancon: Author. Retrieved from http://www.energy-cities.eu/IMG/pdf/internal_contracting_stuttgart_one_pager.pdf.
4. German Watch. (2015). *Finding the finance. financing climate compatible development in cities*. Bonn: Author. Retrieved from https://germanwatch.org/en/download/13426.pdf.
5. Zirkwitz, H.-W. (2016). *Financing the energy transition - Internal contracting*. Besancon: Energy Cities. Retrieved from http://www.energy-cities.eu/IMG/pdf/zirkwitz_intracting.pdf.

6. Seifried, D. (2011). *Finanzierungsmodelle Für Das Kommunale Energiemanagement*. Frankfurt: DUH. Retrieved from http://www.duh.de/fileadmin/user_upload/download/Projektinformation/Kommunaler_Umweltschutz/Klimakommune_2010/Workshops/Workshop_Frankfurt/02_WS-I_Finanzierungsmodelle_Seifried.pdf.
7. Energy Cities. (2017). *Financing energy transition in cities*. Besancon: Author. Retrieved from http://www.energy-cities.eu/Focus-on-INFINITE-SOLUTIONS.
8. Limaye, D. R., & Limaye, E. S. (2010). Scaling up energy efficiency: The case for a super ESCO. *Energy Efficiency, 4*(2), 133–144.
9. ESMAP. (2014). *Financing municipal energy efficiency projects*. Washington, DC: Author. Retrieved from https://www.esmap.org/sites/esmap.org/files/DocumentLibrary/FINAL_MGN1-Municipal%20Financing_KS18-14_web.pdf.
10. European Commission. (2014). *ManagEnergy - Leaders in energy transition*. Brussels: Author. Retrieved from https://ec.europa.eu/easme/en/managenergy?corncr=financial.
11. Publenef. (2017). *City of Litomerice. Revolving energy saving fund Litomerice [Czech Republic]*. Besancon: Author. Retrieved from http://publenef-project.eu/wp-content/uploads/2017/04/GP17_PUBLENEF_factsheet_-Litomerice.pdf.
12. Oxfam. (2017). *Blended finance. What it is, how it works and how it is used*. Oxford: Author. Retrieved from http://www.eurodad.org/files/pdf/58a1e294657ab.pdf.
13. CityInvest. (2015). *Model 10. Energy Fund Den Haag – ED*. Retrieved from http://citynvest.eu/sites/default/files/library-documents/Model%2010_Energy%20Fund%20Den%20Haag_ED_final.pdf.
14. ManagEnergy. (2017). *Municipal bonds emission for energy efficient retrofitting of street lighting – Varna, Bulgaria*. Limerick: Author. Retrieved from www.managenergy.net/instruments/6?casestudy=1505&pagename=usecases#.WMphC_JmqKV.
15. LBNL. (2012). *Using QECBs for street lighting upgrades: Lighting the way to lower energy bills in San Diego*. Berkeley, CA: Author. Retrieved from https://energy.gov/sites/prod/files/2014/06/f16/street-lighting-qecb.pdf.
16. Kinzey, B. (2015). *Restoring Detroit's street lighting system*. Richland, WA: Pacific Northwest National Laboratory. Retrieved from https://energy.gov/sites/prod/files/2015/09/f27/2015_restoring-detroit.pdf.
17. Boza-Kiss, B., Zangheri, P., Bertoldi, P., & Economidou, M. (2017). *Practices and opportunities for energy performance contracting in the public sector in EU Member States, EUR 28602 EN*. Luxembourg: Publications Office of the European Union. ISBN 978-92-79-68832-4. Retrieved from http://publications.jrc.ec.europa.eu/repository/bitstream/JRC106625/kjna28602enn.pdf.
18. Diputacion de Huelva. (2016). *Environment and energy department. Bundling efficient energy investments. An innovative procedure in the province of Huelva*. Huelva: Author. Retrieved from http://laccc.diphuelva.es/galerias/docs/304.pdf.
19. GuarantEE. (2017). *Public street lighting in the City of Tona – Spain*. Berlin: Author. Retrieved from http://guarantee-project.eu/bestpractice/tonas-city-hall-spain/.
20. GuarantEE. (2017). *Street lighting in Sapareva Bania Municipality – Bulgaria*. Berlin: Author. Retrieved from https://guarantee-project.eu/bestpractice/street-lightening-bulgaria/.
21. Geissler, M. (2013). *Energy performance contracting. The Example of Berlin and the EU-wide experiences*. Retrieved from http://slidegur.com/doc/5189263/epc.
22. Sustainable Energy Authority of Ireland. National Energy Services Framework. (2014). *Energy performance related payments (EPRP). Guide*. Dublin 2: SEAI. Retrieved from https://www.seai.ie/resources/publications/Energy-Performance-Related-Payments-EPRP-Guide.pdf.
23. The Climate Group. (2013). *LED street lighting. Climate Group document - Financial section*. London: Author. Retrieved from https://www.theclimategroup.org/sites/default/files/archive/files/LONDON-Working-Document-finance.pdf.
24. Makumbe, P., Weyl, K. D., Eil, A., & Jie, L. (2016). *Proven delivery models for LED public lighting: Synthesis of six case studies* (Energy Sector Management Assistance Program). Washington, DC: World Bank. Retrieved from http://documents.worldbank.org/curated/en/869131477561325418/Proven-delivery-models-for-LED-public-lighting-synthesis-of-six-case-studies.

25. ESCAP. (2008). *A primer to public-private partnerships in infrastructure development*. Bangkok: Author. Retrieved from http://www.unescap.org/ttdw/ppp/ppp_primer/01_why_and_what.html.
26. Mazzolini, M. Director of Agenzia Per l'Energia del Friuli-Venezia Giulia. Feedback to the survey conducted in June 2017.
27. Patrick Duguet. *Head of public lighting of the City of Paris, France*. Email communication on 7 December 2018 and 4 December 2018.
28. De Marco, A., Mangano, G., Michelucci, F. V., & Zenezini, G. (2016). Using the private finance initiative for energy efficiency projects at the urban scale. *International Journal of Energy Sector Management, 10*(1), 99.
29. Esty, B., & Sesia, A. (2010). An overview of project finance and infrastructure finance – 2009 Update. In *Harvard Business Review Background Note 210-061*. Cambridge, MA: Harvard Business School Publishing.
30. Scottish Futures Trust. (2013). *Street lighting toolkit - How to assess the impact of an energy efficiency investment in the street lighting asset*. Edinburgh: Author. Retrieved from http://www.scottishfuturestrust.org.uk/files/publications/Street_Lighting_Toolkit.pdf.
31. WBG, World Bank Group. (2016). *Lighting Brazilian cities: Business models for energy-efficient street lighting*. Washington, DC: Author. Retrieved from https://ppp.worldbank.org/public-private-partnership/library/lighting-brazilian-cities-business-models-for-energy-efficient-street-lighting.
32. Mendoza, E., Gold, M., Carter, P., & Parmar, J. (1999). The sale of highway 407 express toll route: A case study. *The Journal of Structured Finance, 5*(3), 5–14.
33. Spillers, C. (2000). Airport privatizations: Smooth flying or a crash landing? *The Journal of Structured Finance, 5*(4), 41–47.
34. Link, H. (2012). Unbundling, public infrastructure financing and access charge regulation in the German rail sector. *Journal of Rail Transport Planning and Management, 2*(3), 63–71.
35. Bonetti, V., Caselli, S., & Gatti, S. (2010). Offtaking agreements and how they impact the cost of funding for project finance deals: A clinical case study of the Quezon Power Ltd Co. *Review of Financial Economics, 19*(2), 60–71.
36. HM Treasury. (2016). *Private finance initiative and private finance 2 projects: 2016 Summary data*. London: Author. Retrieved from https://www.gov.uk/government/publications/private-finance-initiative-and-private-finance-2-projects-2016-summary-data.
37. European Commission. (2016). *Good practice in energy efficiency. Commission staff working document*. Brussels: Author.
38. Bettervest. *Heating network with two CHPs and retrofitted Heat boiler for Wilhelmstadt gymnasium in Berlin* [Nahwärmenetz mit zwei BHKWs und modernisierten Heizkesseln für das Wilhelmstadt Gymnasium in Berlin]. Retrieved from https://www.bettervest.com/en/projekt/nahwaermenetz_wilhelmstadt_gymnasium_berlin#beschreibung.

Day-Ahead Multi-Objective Energy Optimization of a Smart Building in a Dynamic Pricing Scenario

M. Botticelli, G. Comodi, A. Monteriù, A. Pallante, and S. Pizzuti

Abstract The identification of techniques aimed at a rational use of electric power has nowadays become more important than the production of energy itself. One of the causes for this is the progressive saturation of the Italian electricity grid, which is increasingly subject to connection requests, mainly due to the development of plants which make use of renewable energy sources.

In order to reduce the building's energy costs during the summer season taking into account the user comfort, in this work we propose a new approach based on Pareto multi-objective optimization combined with a simulator developed in the MATLAB/Simulink environment. The electrical consumption of the entire building is taken into consideration with the aim of air-conditioning it. The goal is to find, the day before, the optimal hourly scheduling of the set points which have to be applied the next day, taking into consideration all external conditions, namely the weather conditions and the hourly energy price. To achieve this objective, the control variables we change are the room temperature set points and the flow water temperature set point. As required by the UNI EN ISO 7730:2006 standard (http://store.uni.com/catalogo/index.php/uni-en-iso-7730-2006.html), comfort measurement has been calculated with the PPD (Predicted Percentage of Dissatisfied) index.

Different scenarios have been investigated. The results show that there is an average of 15% potential cost saving, while maintaining a high level of comfort. Experimentation has been carried out by simulating a real office building in Italy, and the comparisons are shown regarding the actual settings applied to it.

M. Botticelli (✉) · G. Comodi · A. Monteriù
Marche Polytechnic University, Ancona, Italy
e-mail: m.botticelli@pm.univpm.it

A. Pallante
Roma Tre University, Rome, Italy

S. Pizzuti
ENEA – Energy, New Technologies and Sustainable Economic Development Agency, Rome, Italy

P. Bertoldi (ed.), *Improving Energy Efficiency in Commercial Buildings and Smart Communities*, Springer Proceedings in Energy,
https://doi.org/10.1007/978-3-030-31459-0_12

1 Introduction

Buildings represent an energetic complex system, made up by a main structure and by its envelope, by air conditioning system and energy vectors which supply it. Only a precise knowledge of subsystems performances and relations among them, allows finding proper strategies for energy saving, maintaining at the same time appropriate comfort levels for residents.

Finding proper techniques for a rational use of electrical energy has become, nowadays, more and more important than the production of energy itself, because the Italian power grid is gradually becoming saturated and being affected by many connection requests coming mostly from plants which use renewable energies.

The present energy consumption which is necessary to the thermal comfort achievement of rooms and of clean water represents almost 30% of the national energetic consumption and it responsible for almost 25% of the national CO_2 emission, which is one of the principal cause of greenhouse effect and temperature heating [1].

An intervention is crucial to reach a new equilibrium, in harmony with environment and respectful of future generation's rights. A change in terms of energetic consumption is necessary [2]:

- for an ethical and social issue: 28% of the world's population currently consumes 77% of the world's energy production, while the other 72% live with only the remaining 23%.
- For a strategic reason: Europe (and Italy in particular) depends on non-EU countries (some highly unstable) for fossil fuel needs and consequently security of supply is not always guaranteed.
- For an economic reason: the annual cost of energy bills is now one of the most important items in the family budget (exceeding the income of a month for an average household) [2].

According to the latest Italian Energy Strategy Plan [3], Rationale Use-of-Energy is considered the cornerstone in order to pursue sustainable economic growth.

Around 40% of total energy consumption is consumed by the building sector. Thus the building sector is also the largest emitter of CO_2 in the European Union (EU). As a result, the conservation and rational use of energy is an important point for European energy policy, as indicated by the European Buildings Directive (EPBD) 2010/31/EU [4, 5], which aims to improve the energy performance of buildings. The member states of Europe have set themselves an energy saving target of 20% by 2020 and 27% by 2030, applying energy efficiency measures [6].

So it is clear that in order to reduce the consumption of electricity, it is necessary to act to reduce the consumption of buildings. The solutions can therefore be two:

- Envelope and plants retrofitting, which leads to a scenario of high savings at the cost of high investments and therefore to a long payback period.
- Automation and intelligence through smart buildings, which leads to medium savings with low investment with a consequent short payback period.

In fact, with the investment in more than ten smart buildings, the payback period is around 3 years, which is much lower than the payback period of retrofitting solutions which is around 15–20 years.

Therefore, monitoring of energy consumption in buildings, especially public buildings, is critical in order to reduce consumption. In a smart building we can control heating, ventilation, air conditioning, lighting, presence, security and other. With sensors, actuators and microchips, we can collect data and manage it according to a business' functions and services. Thanks to appropriate optimization, control and diagnostics algorithms, we can minimize the energy demand of buildings.

This article describes a possible strategy of optimization, suitable for smart buildings, focusing both on the energy consumption reduction and minimization of unhappy residents' percentage caused by thermal "discomfort." This percentage was determined by PMV index (Predicted Mean Vote) and PPD index (Predicted Percentage of Dissatisfied) [7, 8] in the current regulation UNI EN ISO 7730:2006 [9].

According to our study, we can introduce the most advanced concept of Smart Building, conceived as a component of an integrated system that also includes other actors. The energy demand of a building, in fact, is integrated with the dynamic daily price of energy (NUP—National Unique Price) the production of energy through photovoltaic systems and electrical storage. The management of this scenario with several players is managed by an Energy Management System.

This article is an evolution of a previous work [10], that had analyzed, with a previous version of the Simulator [11], the optimization of methane consumption for an entire winter season. The difference of this experimentation is about a different objective function and a single day. In fact, the OF (Objective function) examined in this report takes into account the dynamic price of energy and not the consumption of methane. On the other hand, with a period reduction of simulation, it is possible to set the different set-points from hour to hour making the whole system more dynamic and efficient in view of a hypothetical functioning in real life.

2 Description

Our study is based on the simulation of a real building, located at the ENEA Casaccia Research Centre (Fig. 1). The building was built at the beginning of the 1970s and is characterized by an L-shaped structure distributed over three floors, in which a total of 41 offices of different sizes and occupancy densities (1–2 occupants) are located. The building is oriented to 15° northeast. The air conditioning of each room is controlled by a fan-coil. The heating system consists of a traditional natural gas boiler.

In order to detect the external and internal environmental conditions and the consumption of electrical and thermal energy, an advanced monitoring system has been installed in the building. A MATLAB Simulink simulator based on the HAMbase model [8, 12] was developed to simulate the variables of interest.

Fig. 1 Case study: smart building at ENEA F40

The control variables are:

- Zone set point which can be set using thermostats in each room
- Flow set point which can be set by means of special instrumentation in the building's thermal power plant.

In fact, the variables that can be set in the simulator are actually modifiable in the modeled building.

Simulink was used to model the described building and all its thermal and electrical components and to simulate the behavior of the F40 building the HAMbase software was used, which exploited the internal temperature model, the relative humidity of the indoor air and the energy consumption required for the heating and cooling of a multi-zone building.

The purpose of the software developed is to simulate the energy saving strategies and control logics in order to maximize energy savings, and therefore to minimize the cost. The outputs are:

- environmental comfort;
- euros.

The thermo-hygrometric comfort index returned by the simulator is a subjective condition that depends on a series of factors. The PMV-index (Predicted Mean Vote) predicts the subjective ratings of the environment in a group of people and it is experimentally related to the PPD, an index that represents the percentage of dissatisfied people. The standard UNI EN ISO 7730:2006 [13] establishes the acceptable range of PMV between −0.5 and +0.5, and therefore states that to be in a situation of well-being, the PPD must not exceed 10%. The PPD can be computed using PMV model. Among those taken from the building model, there are four variables, that were set to their typical values—metabolism (70 [W/m^2]), external work (0 [W/m^2]), clothing (1 [–]), and air velocity (0.1 [m/s]).

The simulator also takes into account other sources of heat, and how they affect temperature.

2.1 Description of Optimization Problem

The analysis of the objective function was carried out by changing a decisional variable at a time (zone set-point and delivery set-point), setting it progressively to each single value assumable, respecting the constraints of the problem.

We have a formulation of the problem as follows (Fig. 2):

Where:

- f_1(X1, X2) is the objective function of euro consumption;
- f_2(X1, X2) is the objective function of the PPD;
- X1 is the zone set-point and variation step is 0.5;
- X2 is the flow set-point and variation step is 0.5;
- LB1 is the lower limit of the variable X1 and is equal to 22 °C;
- UB1 is the upper limit of the variable X1 and is equal to 25 °C;
- LB2 is the lower limit of the variable X2 and is equal to 8 °C;
- UB2 is the lower limit of the variable X2 and is equal to 12 °C;

The upper and lower bound limits LB2 and UB2 are the two values between which average delivery temperatures oscillate in Italian public buildings.

So, we have a minimization problem with two objective functions as the consumption in euro [€] and PPD index with two vectors of 24 elements, X1 (the zone set-point that change from 22 to 25 °C) and X2 (the flow set-point that change from 8 to 12 °C). X1 and X2 are the two control variables.

3 Experimentation

Each simulation is about a single day for which it is possible to set the different set-points from hour to hour. This made the system dynamic and efficient in the vision of a hypothetical functioning in reality.

The first step was to integrate the simulator with the optimizer that is a genetic optimizer (NSGA-2) [12].

Fig. 2 Formulation of problem

Min $f_1(X1, X2)\,, \quad f_2(X1, X2)$

Subject to :

$$\begin{cases} LB_1 \leq X1 \leq UB_1 \\ LB_2 \leq X2 \leq UB_2 \end{cases}$$

The simulator takes in as input both, the values of the decision variables that we have just set up and the final states of the simulator returned by the optimization system.

The goal of the problem is to minimize a multi objective function with cost and comfort (euro [€] and PPD [%]).

For this type of algorithm, the fundamental parameters have been chosen and set, such as:

- Population = 8
- Generation = 4
- Crossover rate = 0.8
- Mutation rate = 0.5

These parameters have been chosen in order to obtain a Pareto front that can be easily reconstructed for each hour. The simulations, however, were also carried out by reversing the number of population and generation.

Crossover and mutation is useful to obtain new solutions that previously did not exist by combining or changing the bits of two solutions.

The total number of fitness evaluations for the exhaustive search is 63 while with the proposed approach is 40. 40 is a reasonable number not to cover all the exhaustive research (63 evaluations).

40 = 8 (pop.) × 4 (Gen) + 8 (pop)

Therefore, in the proposed approach the number of performance evaluations is fewer. In fact, the iteration of the algorithm are lower (40) than the exhaustive search (63).

In the problem's solution for minimizing cost consumption, expressed in terms of euro, and PPD, is dynamically calculating the values of the set-points in the totality of their combinations, using the genetic evolutionary algorithm NSGA-2. Solutions are ordered according to the concept of non-dominance (Fig. 3).

The population of individuals is random initialized and the optimization process begins. When the population is initialized, the chromosome phenotype associated with each individual is calculated and, subsequently, the fitness of each individual is calculated. If the stop condition of the algorithm is verified the algorithm ends, otherwise the genetic selection operators are applied, crossover and mutation. The count of the number of generations carried out is increased and the new population, which replaces the previous one, is evaluated again.

As we have already said for the hours following the first one, the optimal solution of the previous hour is used as initialization for the next hour.

For the choice of the best solution, the logic used is therefore to order the Pareto front based on increasing euro consumption and scroll through this list checking the PPD. The first value just below 10% is the chosen one. If all solutions have a PPD higher than 10, then the best solution is the chosen one with the smallest possible PPD value.

The daily simulations were carried out for an average day in summer.

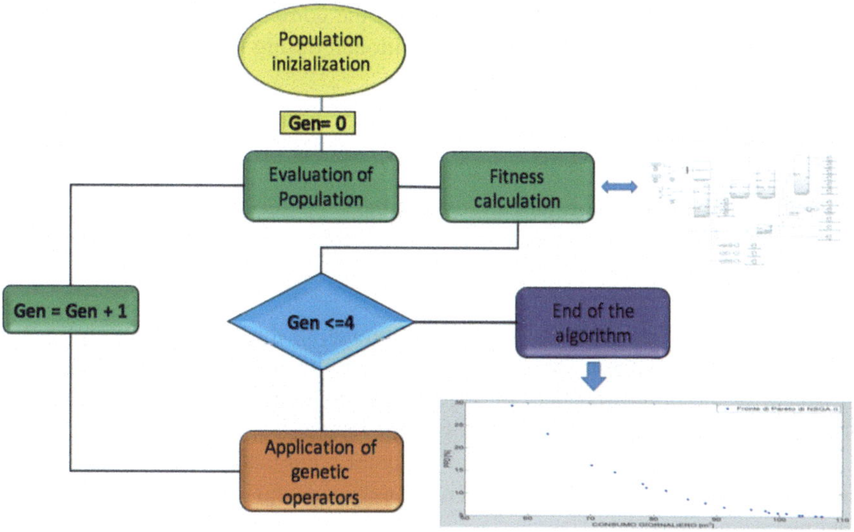

Fig. 3 Problem's solution

Fig. 4 Real functioning of the building in comparison with the optimized one

To evaluate the efficiency of the algorithm, the combination of set-points with which the F40 building is currently air-conditioned during the summer season has been taken into consideration:

- Zone set-point = 22 °C;
- Delivery set-point = 10 °C;

which have been kept constant throughout the simulation period (Fig. 4).

The simulations took place by taking three different initial internal temperatures:

T0 = 21; 23; 25 [°C]

We can observe the results obtained by simulating the real functioning of the building (Table 1), without any kind of optimization, with static set-point, in comparison with the optimized one, in which each simulation involved changing the flow set-point and the zone set-point in specific ranges.

For the identification of the optimal set-point combination with NSGA-2, the same initial internal temperatures were considered and the different solutions obtained were compared.

It can be seen from the results in Table 2, on average we found a "theoretical" saving of 14.09% compared to the reference case, with an average PPD value of 8.76.

We talk about "theoretical savings" because the strategy has not been applied to the real building.

We can see in Table 2 that 10.89 is above the threshold of 10% of dissatisfied, in fact starting already with a higher temperature the dissatisfied are already many and it is therefore significant the fact that compared to a very small increase of dissatisfied (10.82–10.89) the savings in terms of euros is substantial, 11.36%.

In the first graph below (Fig. 5) we can see how the punctual consumption expressed in euros varies in the 24 h, comparing the one of the reference case (in blue: Setpoint zona = 22 °C; Setpoint di consegna = 10 °C) with the consumption obtained using the NSGA-II optimizer (in red).

We can therefore note that savings were obtained at all hours of the day, obtaining an overall saving equal to the difference between the integrals related to the two curves.

In the second graph below we can see how the punctual PPD value varies in the 24 h, comparing the one of the reference case (in blue) with the consumption obtained using the NSGA-II optimizer (in red).

It is clear that in the face of savings, we are going to lose the value of PPD, but it is always below 10% as defined by the UNI standard.

Table 1 Reference case (Summer, one average day)

Initial temperature	21 °C	23 °C	25 °C
Euro consumption [€]	22.74	29.84	36.84
PPD [%]	6.17	6.74	10.82

Table 2 Result obtained

Initial temperature	21 °C	23 °C	25 °C	Average
Saving [%] NSGA-2	−9.76	−21.15	−11.36	−14.09
PPD (average) NSGA-2	7.13	8.25	10.89	8.76

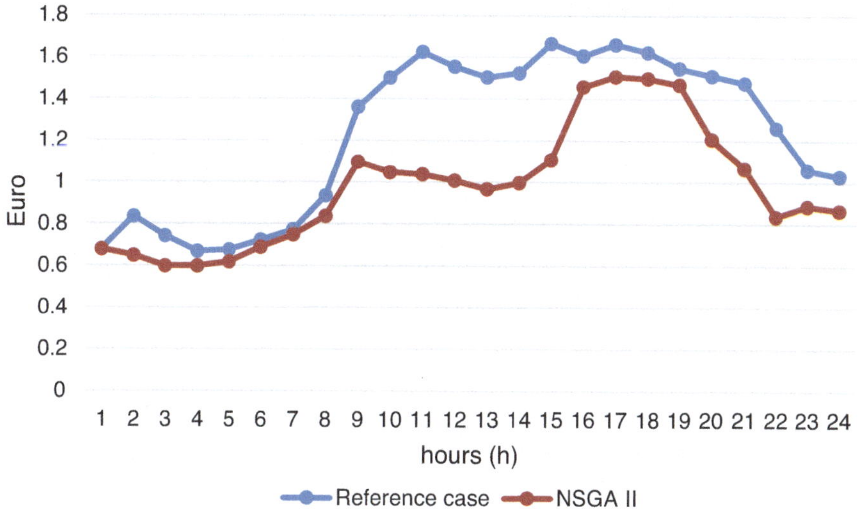

Fig. 5 Punctual euro consumption in 24 h

So with a small increase of PPD value, but always respecting the legislation, we have however a great cost saving (Fig. 6).

4 Conclusion

In this work we proposed an approach based on multi-objective optimization combined with simulation tools in order to improve the building thermal energy performance during summer by taking into account euro consumption and occupant comfort. Results were carried out by simulating the behavior of a real building on a one average day in summer. In particular different scenarios were investigated and results showed that there is a nearly 15% average potential of energy saving with little increase of user discomfort through an hour-by-hour multi-objective optimization.

Future developments include firstly the comparison of this results by using other optimization methods. Then it will be introduced a new version of the simulator that integrates PV (photovoltaic) and storage for even more efficient management of energy consumption. To obtain a more complete scenario, simulations will have to

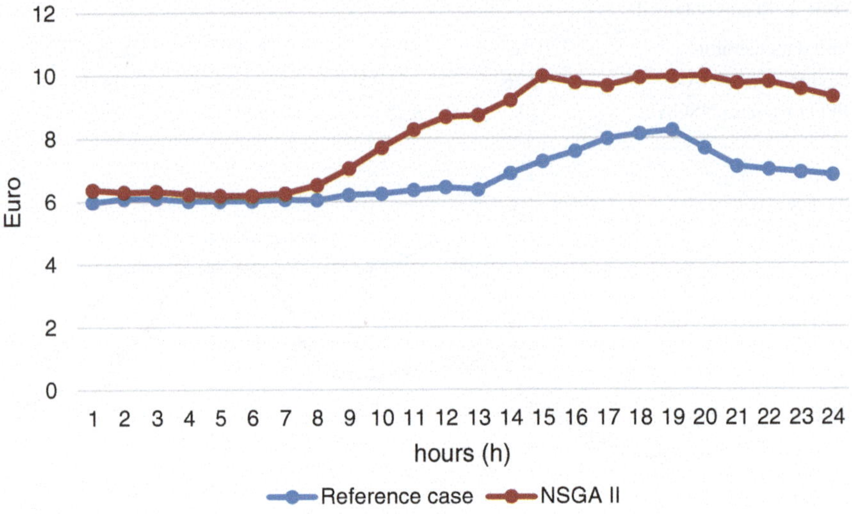

Fig. 6 Punctual PPD value in 24 h

be carried out covering the winter season. Finally, to validate the solutions proposed in a real case, the application of this strategy to the smart building F40 is envisaged. Further future development could consist in adapting the work done to residential buildings.

References

1. ENEA. (2003). *Risparmi energetico nelle case*. Rome: Author. Retrieved from http://efficienzaenergetica.acs.enea.it/doc/risparmio_casa_agg.pdf.
2. Isolani, P. (2008). *L'uso razionale dell'energia negli edifici pubblici*. Retrieved from https://www.google.it/url?sa=t&rct=j&q=&esrc=s&source=web&cd=1&ved=0ahUKEwi1y_L0rvDXAhWEbRQKHV4CCqMQFggrMAA&url=http%3A%2F%2Fwww.ri.camcom.it%2Fdownload%2F775.html&usg=AOvVaw0LNQCreHWzF9Lb0RoqZJbA.
3. Italian Energy Strategy Plan. Retrieved from http://www.sviluppoeconomico.gov.it/index.php/en/news/2037432-national-energy-strategy.
4. European Union. (2003). Directive 2002/91/EC of the European Parliament and of the Council of 16 December 2002 on the energy performance of buildings. *Official Journal of the European Union*. L1/65.
5. European Union. (2010). Directive 2010/31/EU of the European Parliament and of the Council of 19 May 2010 on the energy performance of buildings (recast). *Official Journal of the European Union*. L153/13.
6. Hyvarinen, J., & Karki, S. (1996). *Building optimization and fault diagnosis source book, IEA Annex*. Espoo: VTT Technical Research Centre of Finland.
7. Bovesecchi, G. *Comfort termo-igrometrico: Equazione del comfort, PPD e PVM*. Rome: Università di r Roma Tor Vergata. Retrieved from http://didattica.uniroma2.it/assets/uploads/corsi/144485/04_-_TT1_-_Comfort_Termoigrometrico_-_Fanger_PPD_e_PMV1.pdf.

8. Fanger, P. O. (1970). *Thermal comfort: Analysis and applications in environmental engineering*. New York, NY: McGraw-Hill.
9. Normativa UNI EN ISO 7730:2006. (2006). *Ergonomics of the thermal environment-Analytical determination and interpretation of thermal comfort using calculation of the PMV and PPD indices and local thermal comfort criteria.*
10. Camponeschi, M., Fonti, A., Leccese, F., Comodi, G., Grossoni, M., & Pizzuti, S. (2015). Winter thermal multi-objective optimization: A simulation case study. *International Journal of Engineering Science and Innovative Technology (IJESIT), 4*, 1. Retrieved from http://hdl.handle.net/11590/283981.
11. Comodi, G., Fonti, A., Giantomassi, A., Polonara, F., & Longhi, S. (2013). *Sviluppo di un simulatore di edifici orientato alla gestione attiva della domanda*. Rome: ENEA.
12. Deb, K., Pratap, A., Agarwal, S., & Meyarivan, T. (2002). A fast and elitist multiobjective genetic algorithm: NSGA-II. *IEEE Transactions on Evolutionary Computation, 6*(2), 182. https://doi.org/10.1109/4235.996017.
13. UNI EN ISO 7730:2006 standard. Retrieved from http://store.uni.com/catalogo/index.php/uni-en-iso-7730-2006.html.

Approach and Decision-Making Process for Sustainable Retrofitting of Commercial Buildings

Enrique Grosser and Kinga Horváth

Abstract The main stakeholders in a retrofitting process for public and commercial buildings are the Lessor, Users, Investors, and Developers. They all have different market and personal interests, which need to be taken into account when developing optimal, affordable, and successful projects. These stakeholders will consider technical, energy, legal, and economic aspects of a holistic retrofitting solution that fit into their interest chain both from an individual and from their shared perspective. Therefore, a value-driven process has to be set up to underline the whole retrofitting project.

In practice, stakeholders need to define and clearly state a common goal and retrofitting strategy, as well as to set up a shared management to design and implement the project. Behind the retrofitting strategy, there needs to be a business case for the project that should include not only energy savings but other factors, so-called co-impacts, such as the increased market value of the property.

The main drivers of retrofitting public and commercial buildings typically include cost savings and improvement of the building state. Typical technical solutions are energy efficiency improvement, reduction of energy demand, and reduction of building related emissions. However, it is no longer possible to implement a retrofitting project without considering the comfort level determined by indoor air quality and thermal comfort, but also by the functional, aesthetical and environmental factors.

The implementation process developed by the EcoShopping* project can be considered as a model for achieving such a holistic retrofit implementation.

This project has been funded with support from the European Commission. This publication reflects the views only of the author, and the Commission cannot be held responsible for any use which may be made of the information contained therein.

E. Grosser (✉) · K. Horváth
EnergoSys Inc., Budapest, Hungary
e-mail: enrique.grosser@energosys.eu

P. Bertoldi (ed.), *Improving Energy Efficiency in Commercial Buildings and Smart Communities*, Springer Proceedings in Energy,
https://doi.org/10.1007/978-3-030-31459-0_13

1 Retrofitting of Commercial Buildings

Buildings account for 40% of energy consumption in the European Union [1]. Energy-efficiency improvement targets for the building stock can play a significant role in the decarbonization goals of the European Union. By 2020, the European Union commits to reduce overall greenhouse gas emissions by at least 20% below 1990 levels, and by 30% in the event of an international agreement [1]. Targeting existing buildings offers the most savings potential, since buildings account for 40% of total energy consumption in the Union, but their renovation rate is very low at around 0.4% to 1.2% per year [2].

Commercial buildings are buildings that are used for commercial purposes, and include office buildings, warehouses, and retail buildings (convenience stores, "big box" stores, and shopping malls). In this paper, we exclude office buildings and target only the other, shopping type buildings.

Shopping buildings account for 43 TW h in 2012, for their needs of space heating, cooling, appliances, ventilation, refrigeration, and lighting [3]. This paper reviews their specific challenges for energy-efficient retrofitting, introducing a holistic approach that was developed during the research project called "EcoShopping" [4], as a way to overcome several of the challenges.

Energy costs are in the biggest cost of a typical commercial building's operating expense, and they are controllable; thus, it gives excellent opportunity for savings. Most commercial buildings use more energy than necessary. The goal is to develop models that achieve energy reduction within market conditions, considering the many factors of building retrofitting processes.

2 Development of a Holistic Approach

From 2013 to 2017, a group of scientists, engineers, economists, and specialists from 16 companies carried out the EcoShopping project. The project aimed to build a holistic retrofitting solution for commercial buildings to reduce primary energy consumption to less than 80 kW h/m^2 per year and to increase the share of renewable energy sources by 50% compared to the baseline. In association with the project, the holistic approach means that the complete refurbishment process aims to provide solutions to shopping centres that equally consider technical, economic, architectural, and social aspects.

The intention of the project was to use and integrate market-available products and technologies, while also developing new technologies. It aimed to reduce energy consumption meanwhile improving users' comfort. The project provided high energy performance, high security, and improved health conditions in commercial buildings. These aims have been achieved with reliable services to visitors, building operators and owners, with automatized processes.

The approach developed by the EcoShopping does not only focus on retrofitting technologies but also at best practices for retrofitting and implementation procedures. The project developed a methodology for holistic retrofitting which is summarized in a public user guideline. Furthermore, support tools were also developed, including an insulation and self-cleaning materials database and an improved life-cycle analysis method, which was extended beyond materials and technologies to economic life-cycle analysis. Also, during the project, recommendations have been made for improved implementation procedures based on the experiences of a demonstration project (Fig. 1).

The guideline for holistic retrofitting provided by the EcoShopping project can be downloaded from: http://ecoshopping-project.eu/document/D2.11_Methodology%20for%20holistic%20retrofitting%20in%20commercial%20buildings_V7_final_with%20apppendixes.pdf

3 The Retrofit Process

The implementation process developed by the EcoShopping project is a model for achieving a holistic building retrofit implementation.

The approach consists of consecutive steps aimed at retrofitting assessment, definition, and deployment:

- Project setup and pre-retrofit survey to assess existing conditions
- Energy audit and performance assessment

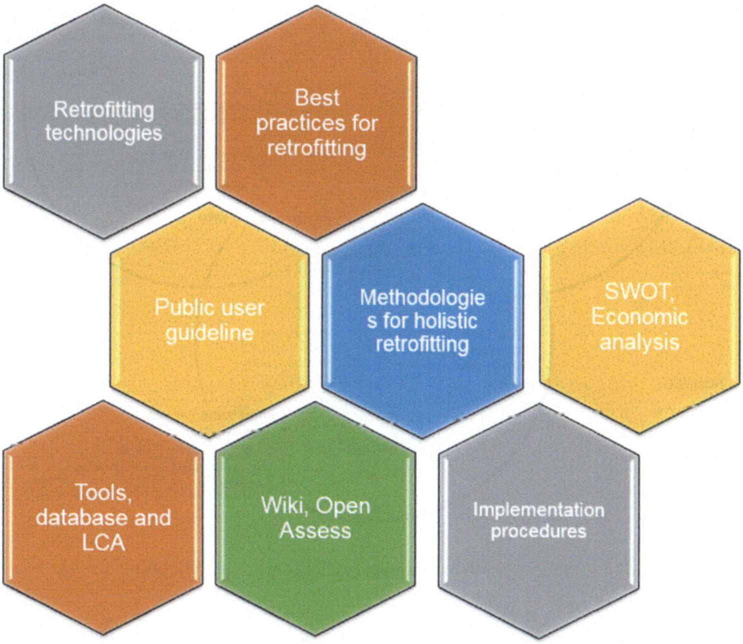

Fig. 1 The components of the EcoShopping project

- Identification of retrofit options and feasibility study
- Implementation of energy-efficiency measures on the site and testing
- Validation and verification of the measures and the implementation.

There are many influencing factors in this process, such as outdoor environmental conditions, regulations, technological options, socioeconomic characteristics, market, energy tariff, O&M, energy systems, and renewable energy options.

The main drivers for a retrofit are usually cost saving and improvement of the building's condition; among the solutions, the energy efficiency improvement, reduction in energy demand and emission of the building are the main areas. However, it is no longer possible to implement a retrofit project without considering the comfort level which is determined by an air quality and thermal comfort conditions, but also by the functional, aesthetical, and environmental factors.

Figure 2 shows the areas, aspects, and influencing factors of creating a holistic design strategy, focusing on a project (building) based approach.

The technical implementation strategy always needs to be complemented with a business case for the project, including not only the energy savings but also the increase of market value of the building and the improved services, to reflect the actual market value of the retrofit. The holistic retrofitting should be value-oriented.

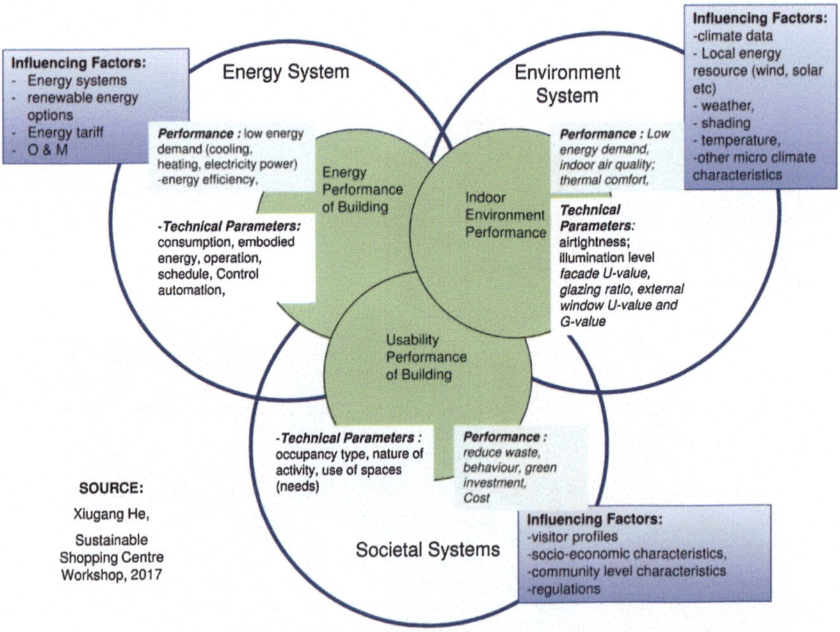

Fig. 2 Holistic design strategy (Source: Xiugang He, Sustainable Shopping Centre Workshop, Budapest, 2017 [5])

3.1 Stakeholders

The main stakeholders in a commercial building retrofit project are the Lessors, Users, Investors, and Developers. They all have different market and personal interests. The Investors are primarily income-oriented and the Developers are profit-oriented. The lessor are cost-oriented, they do not want to pay high fees. End users want comfort and high level of services.

But why does not the retrofit happen then, and what are the barriers in the interest chain? We found that the aspects of sustainability only potentially motivate market stakeholders to build or retrofit energy-efficient commercial buildings. Every stakeholder aims to lower their own risks.

The common interest of the variously oriented stakeholders is that the investment shall be cost-effective, and the economic, functional, operational, and environmental benefits shall deliver the expected return confidently.

This is only realistic if the results can be guaranteed. Previous business models that include a financial guarantee, such as an Energy Performance Contracting or ESCO model do not resolve the blockings in the interest chain.

3.2 Value-Oriented Energy-Efficient Retrofitting

The art of developing an optimal, affordable, and successful retrofit project depends on the capacity to integrate the above described different economic interests of all of the involved stakeholders. In order to reach the goals of the European Union, stakeholders need to have economic interest in energy saving.

Stakeholders do not have the same incentives for energy savings. They need individually and jointly consider technical, energy, legal, and economic aspects forming a holistic retrofitting solution that fits their common interest chain. The project implementers have to set up a value-driven process.

After doing so, the stakeholders also need to define a common goal, formulate a retrofitting strategy and set up a common management to design and implement the project, which goals need to be agreed and clearly defined (Fig. 3).

We found that to comply with the above requirements of differing interest, the model should be:

- VALUE CREATOR
- COMPETITIVE
- ADAPTABLE
- FUNDABLE

The question arises, whether sustainability and high energy performance can increase the market value of shopping centres.

- It can lower the Gross Operative Costs, in particular energy, operation, maintenance, and monitoring costs.

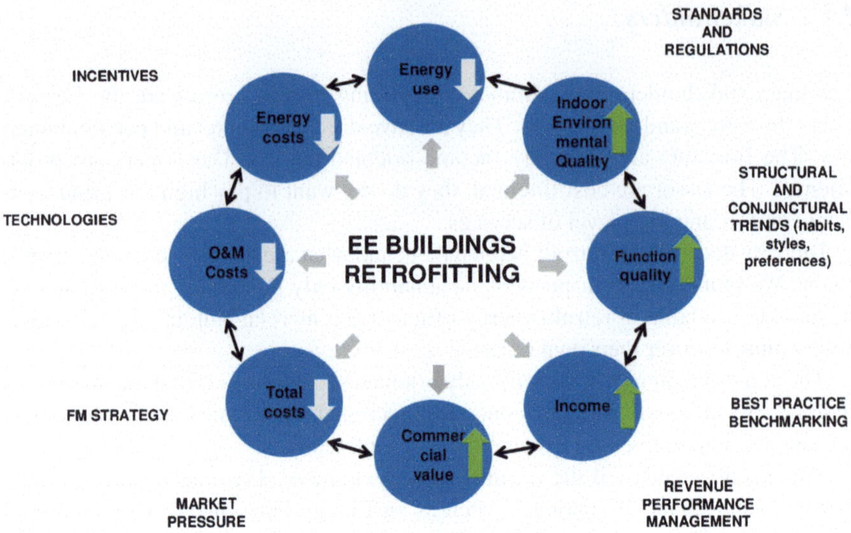

Fig. 3 Impact of value-oriented retrofitting on commercial buildings (Enrique Grosser, Kinga Horváth, Sustainable Shopping Centre Workshop, 2017 [6])

- It can increase the indoor environmental quality, and consequently the performance and health of users/workers.
- It can increase the well-being of visitors and consequently the time they spend in the facility.
- And by these, it can increase the attractiveness and traffic of the shopping center, and achieve a better market position of the building, an increase of rental fees and higher utilization of its capacity.
- Consequently, the gross operative yield of the shopping centre can increase, and in relation to that, its overall market value can increase.

Together with the economic aspects, the retrofitting process needs to follow an integrated approach on a social or individual level. Key factors of the integrated approach:

1. Common goals:
    (a) User satisfaction;
    (b) Sustainable rentability;
    (c) Acceptable return;
    (d) Positive image.
2. Common management.
3. Pillars of efficiency:
    (a) The measurement, verification, evaluation, and design systems of the energetic and functional processes.

(b) The quality and adequateness of the technologies, the preparation and realization of the investment.
(c) Monitoring and control of the energy systems and processes.

4. Continuous feedback and adaptation, turning the process into an approach that considers the whole life cycle of the building portfolio.

Consequently, a two-phase assessment method is suggested here to be followed for retrofit investments:

1. Strategic positioning: a commercial assessment of the future competitive position of the commercial facility, including catchment area study, visitors' profile, tenant mix, expected rent, and exit yield. Competitiveness is the key factor for investors to be interested in investing in these kinds of projects.
2. Detailed redesign of building structure, energy (and other sustainable component) saving targets, costing and the establishment of new operational regime.

Figure 4 shows the circular revolving process of managing a commercial portfolio based on the integrated approach that combines the aspects described in this chapter.

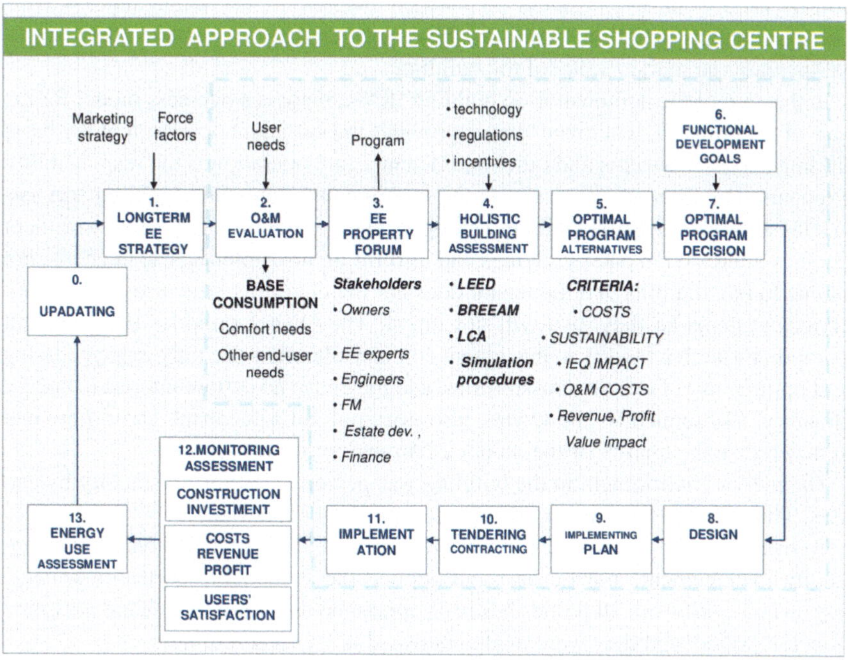

Fig. 4 Integrated approach to the sustainable shopping centre (Enrique Grosser, Kinga Horváth, Sustainable Shopping Centre Workshop, 2017 [6])

4 A Demonstration

We developed the EcoShopping solution further and tested it during a demonstration case, where we have used the EcoShopping technologies and methods. The demonstration case has been the energy-efficient retrofitting of a shopping centre in Hungary. The shopping centre is called "IKVA," and it had not been renovated prior to the demonstration, since its building in 1979, so it has been in a bad condition. The building has approx. 2000 m^2 leasable area. During the development of the project and the business case, the value-oriented holistic retrofitting approach has been used as a guideline. Now, we would not go into technological details, but rather describe the unique factors that mostly affected how the project has been formed in order to provide value for the stakeholders.

Real estate developers want to increase their net operating income, like the Owner of the IKVA Shopping Centre, who has an extensive portfolio consisting of buildings with various commercial profiles. It became evident that we cannot encourage market stakeholders to develop an urge to save the planet if it endangers their business aim. Consequently, retrofitting projects will only happen if they ensure the expected profit from the investment.

In practice, stakeholders of projects aiming to provide energy efficiency have to develop common energy efficiency goals and a strategy to achieve them. They need to build a common team to develop and execute the project. Beyond this strategy, projects need to have a business plan which ensures the economic success. This business plan has to not only consider the energy savings but also the other factors that increase the market value of the building.

In the case of the demonstration building IKVA, there is a business model for the lease of the building, where the Owner provides the sufficient operation of the basic building services systems and other maintenance and operation is the responsibility of tenants. However, in this case, as the building structures and building services systems and the operation of them belong to two separate parties, neither of them becomes interested in energy savings and retrofitting investments. The Owner is not the one to pay the bills and the tenant does not own building structures or building services systems so they also will not invest. The Owner provides the sufficient operation through a facility management company, but they are only engaged to the most optimal use of existing possibilities, and they have no certain interest in energy efficiency. Consequently, the Owner also does not get a feedback about how end users perceive the quality of the indoor environment of the building.

The indoor environment in the building was perceived as not satisfactory by end users. This affected the market position and the level of services.

To overcome the obstacles, we have worked with all three stakeholders (Owner, Facility management, and Leaser) and applied the sustainable portfolio strategy. Responsible staff from all parties has been appointed and all of them were informed about the project and they were made interested in its successfulness.

Engineers, technological, economic, project development, and execution experts developed multiple technological, financial, and organizational scenarios. All scenarios took the market repositioning of the building into account. The EcoShopping consortium provided a solution applicable in each scenario. The level of intervention depended on the decision of the Owner who decided not just upon the viability of the solution but considering the current status of his portfolio.

Stakeholders decided to carry out a sequenced project, where the technologies are implemented according to a long-term plan. The Owner and the Leaser agreed on a new lease agreement, and the tenant understood that energy-efficiency measures not only lower their fees but also improve the provided services. The selected solution has been implemented and evaluated within the framework of the EcoShopping demonstration project. Two factors were prominent: costs were lowered by technological solutions, and this also made possible that the renewed building services systems provided better indoor quality.

After 2 years of monitoring the results, we can say that the project reached the desired level of energy savings and provided a renewed quality of services for end users. As a result, the tenant, who previously considered leaving the building, changed their approach and even leased the previously empty areas of the building, and their market position changed and their business became more rentable, even with increased rent fees.

The optimal strategy and technological integration provided better conditions and cooperation within stakeholders, who realized that they all can gain from the renovation. There was an interest conflict and lack of interest before the project, therefore the market position became lower and lower. But stakeholders managed to integrate their interests and use the demonstration project to better their position.

5 Challenges

We encourage building retrofit implementers to follow the "case study" approach, encouraging the researchers and developers to apply their collective knowhow to assess the practical possibilities for retrofitting of older, perhaps already "in distresses" commercial buildings. It is necessary to include commercial knowhow into energy-efficient processes development. Involvement of and networking with the actors and stakeholders of EE services for Commercial Buildings are found to be essential elements of a retrofit project.

Overall, we have identified three key goals that should be followed through and kept in mind during an energy-efficient retrofit project of shopping centres in order to make them successful, namely:

1. sustainability;
2. human health and comfort;
3. social and economic value of the building and the related services.

6 Conclusions

It was found in the EcoShopping project that considering market conditions a deep energy or climate retrofit can only produce sufficient return on the investment if the commercial value increases. However, such a commercial/market value cannot and should not be expected to come only from the energy costs savings. The value of the retrofit implementation should be considered as a combined effect of functionality, indoor environmental quality, and others (transport, accessibility, aesthetic, etc.) in addition. Linking market repositioning and deep energy-efficient retrofitting is the key factor for the future of sustainable commercial buildings stock.

References

1. DIRECTIVE 2010/31/EU OF THE EUROPEAN PARLIAMENT AND OF THE COUNCIL of 19 May 2010 on the energy performance of buildings
2. European Commission (2016), An EU Strategy on Heating and Cooling (COM (2016) 15 final), available at: http://eur-lex.europa.eu/legalcontent/EN/TXT/PDF/?uri=CELEX:52016DC0051&rid=1
3. http://eeg.tuwien.ac.at/commonenergy/conclusions, download date: 20-09-2018.
4. See more at: www.ecoshopping-project.eu. The project's full title is EcoShopping Energy efficient & Cost competitive retrofitting solutions for Shopping buildings. This project has been funded with support from the European Commission. This publication reflects the views only of the author, and the Commission cannot be held responsible for any use which may be made of the information contained therein.
5. He X. (2017). Paper presented at the Sustainable Shopping Centre Workshop, Budapest, 2017. http://eeg.tuwien.ac.at/commonenergy/conclusions. Download date: 20-09-2018.
6. Grosser E., & Horváth, K. (2017). Paper presented at the Sustainable Shopping Centre Workshop, 2017.

Building Efficiency Models and the Optimization of the District Heating Network for Low-Carbon Transition Cities

Guglielmina Mutani, Valeria Todeschi, Elisa Guelpa, and Vittorio Verda

Abstract Nowadays, greenhouse gas emissions continue to increase with the consequent climate changes. Energy consumption of buildings strongly affects atmospheric pollution, therefore for a sustainable development it is necessary to adopt energy efficiency policies combined with low-carbon technologies. In particular, the use of district heating (DH) has environmental and economic advantages in energy production and distribution for space heating consumption. In this paper, the combined effect of DH expansion with different buildings retrofit scenarios using a GIS-based model is proposed for a more sustainable city.

This methodology is applied to the DH network of the city of Torino and, energy savings hypotheses were analyzed, evaluating different energy saving trends starting from the current one with existing policies. A GIS-based methodology has been developed with bottom-up and top-down approaches; then two future energy savings scenarios have been hypothesized. Energy retrofit measures have been applied to the most critical areas with low potential of heat distribution; in a second phase, to the whole area connected to the DH network. The results showed that intervening in the critical areas only +5% of potential buildings can be connected to the existing DH network (standard retrofit) while this percentage could grow up to +25% with advanced buildings retrofit. On the other hand, intervening on the whole city, there is a considerable reduction of consumptions and the connectable quota of buildings to the DH network reaches +42% with standard retrofit and +82% with advanced retrofit scenario with an optimization of energy distribution as well.

Relevant topic: smart and sustainable districts, communities and cities.

G. Mutani (✉) · V. Todeschi · E. Guelpa · V. Verda
Department of Energy, Politecnico di Torino, Torino, Italy
e-mail: guglielmina.mutani@polito.it

P. Bertoldi (ed.), *Improving Energy Efficiency in Commercial Buildings and Smart Communities*, Springer Proceedings in Energy,
https://doi.org/10.1007/978-3-030-31459-0_14

1 Introduction

Sustainability and environmental issues are closely associated with energy consumptions and CO_2 emissions that contribute to global warming and air pollution [1]. Climate change is a global phenomenon that largely impacts the quality of urban life and livability. Cities are a key contributor to climate change, as urban activities are a major source of greenhouse gas (GHG) emissions. They are responsible for 75% of global CO_2 emissions, with transport and buildings being among the largest contributors. The improvement of the energy efficiency (EE) is essential to mitigate the GHG emissions, including enhanced air quality, increased social welfare, market competitiveness, and creation of jobs [2]. The use of renewable energy and an increase in EE are two essential solutions to address the energy crisis. Considering the long life of buildings and the low demolition rate that characterize European building stocks, the considerable retrofit of buildings is crucial to reach successful energy saving goals [3]. Most of the building stocks in Mediterranean cities have too high energy consumption and, the Italian building stock is quite old and not adequately renovated. Data of residential buildings shows that 44% of them were built before 1961, 45% of buildings were built between 1961 and 1991, and 11% between 1991 and 2011 [4]. In order to decrease the energy consumption in the existing buildings, a diffused and effective retrofit strategy is needed. The actions of the local authorities and the effectiveness of municipal energy action plans play a crucial role. The existing European policies should be able to improve the energy performance of this housing stock promoting EE measures [5].

District heating (DH) systems can play a significant role in achieving sustainability in building related energy production and distribution [6]. DH systems are useful to reduce the impact on the environment (GHG emissions) and to decrease the fuel demands using the cogeneration combined with other efficient technologies [6–8]. Information about the existing building stock, cost-benefit analysis of EE renovation, possible reductions in energy use and socioeconomic impacts on territory at different scales (regions, cities, districts and neighborhoods) are important to facilitate policy-makers in identifying appropriate strategies and to support stakeholders [9]. The feasibility of retrofit measures in Turin, with the existing energy policies, depends mainly on: the period of construction that affects the need to renovate a building; the level of use of the building (percentage of occupied apartments), the inhabitants' educational level, the per capita income, and the employment rate [10].

To improve the EE of buildings it is possible to use different solutions considering the characteristics of the urban context. Individual condensing boilers represents the best solution at the single building level, but at urban level, the connection to the DH network is more advantageous due to the more efficient technological mix [3].

In most developed countries, heating, cooling, and electricity are available in all buildings, and cost is often a driver to determine if replacing individual end-user systems to a DH system is a viable solution. In low-density context, DH system may not offer any economic advantage, due to the high costs [6, 11, 12]. In fact, DH tend to be more economic for higher density and more populated regions, and governments can facilitate greater use of the DH technology through appropriate incentives [13].

This paper evaluates the implementation of energy policies at urban level to optimize the energy demand and supply of buildings. This analysis considers the buildings served by the DH network, the space heating consumptions (buildings' energy consumption depends on many factors including weather conditions, users' behavior, and buildings' characteristics) and the performance of the network through the use of a fluid dynamic model [14, 15].

The aim of this paper is to develop an optimization assessment at urban level with a sustainable perspective to help decision makers in performing energy planning, through the evaluation of the real buildings' heritage, the use of the DH network and future trends of building retrofit measures.

The research starts from an analysis of the maximum quota of buildings connectable to the existing DH network with the existent transport pipelines not modified. This maximum quota of buildings connected to the DH network depends on territorial, economic and technical limits; technical limits consider the maximum flow rates or velocities into the pipelines. Energy consumption of buildings was evaluated by a GIS-based simplified model derived by the comparison of top-down and bottom-up models for space heating energy consumption of residential and non-residential buildings [15, 16]. The GIS-based models allowed to define, with a good spatial accuracy, the energy demand and supply location while the simplified models did not required a detailed description of building geometry, thermal properties of the building envelope and heating system [17]. The bottom-up models were also useful to identify buildings that require urgent retrofit interventions and to quantify the energy savings after retrofit measures. Figure 1 shows the methodology used in this study.

In the presented research, the principal data consist in characteristics of:

- Turin's buildings: the available data on the building stock provided by the Municipality and the official statistical institution, such as number of buildings, period of construction, surface to volume ratio *S/V*, heated volume, information about the type of user (residential, tertiary, municipal, and industrial), occupancy, and central/individual heating systems. The outdoor microclimatic conditions and mainly the HDD at 20 °C, registered by ARPA weather stations, have been associated to the nearer buildings.
- Turin's district heating network: the local district heating company provided space heating consumption data and the heated buildings volume for 36 meshes of the size of 1 km × 1 km around the existing DH network for the city of Turin. The energy-use data have a monthly detail, for three consecutive heating seasons: 2012/2013, 2013/2014 and 2014/2015.
- Turin building retrofit interventions: the available data on building retrofit were derived from the energy performance certificates (EPC) registered in Piedmont Region from 2009 to 2015 [18]. The EPC database was created considering 103,742 apartment buildings, of which only 4333 are for retrofit interventions (energy efficiency renovations and installation of renewable energy technologies), and 1841 detached houses, of which only 285 are for retrofit interventions.

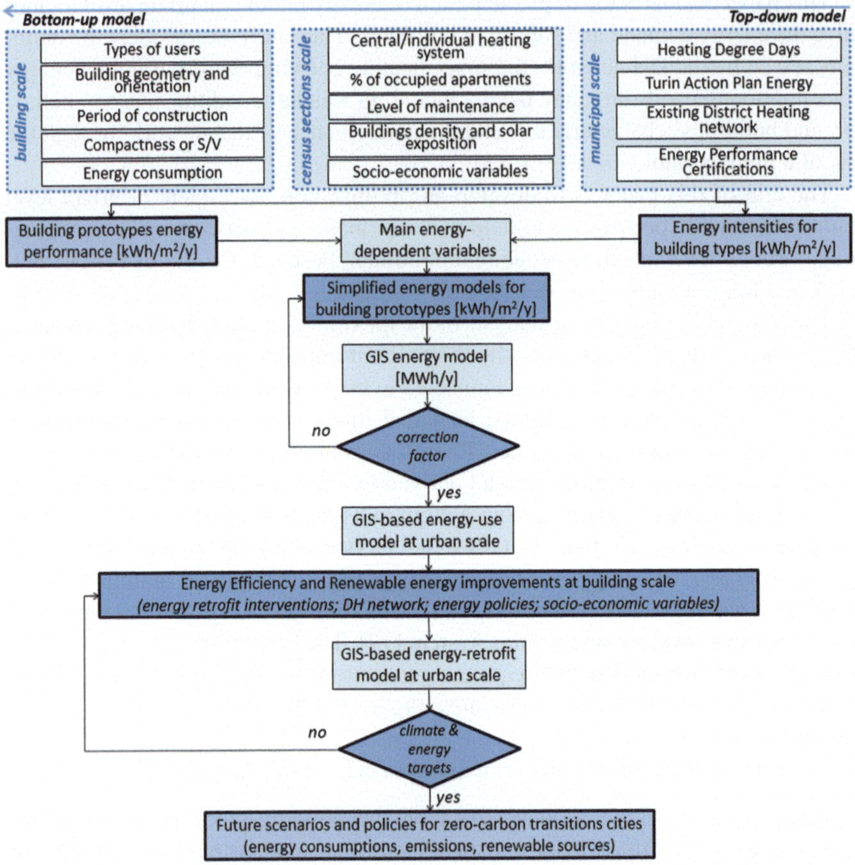

Fig. 1 Flow chart of the methodology approach: energy efficiency model

With the support of Geographical Information Systems (GIS) tool, an updated database of the buildings stock in Turin was developed with all the available georeferenced information (at different scales). The GIS has the advantage to localize and to spatially describe all information to visualize, analyse and plan the energy demand and supply of buildings at local and territorial scale [9]. The use of GIS tool supports the identification of particular buildings or clusters of buildings, with a qualitative and quantitative information on their energy performance. With these tools it is possible to identify the most promising areas to be involved in the process of energy renovation and in simulating possible energy savings scenarios [19]. The final GIS database developed in this study contains: the typological and dimensional data of buildings; the buildings spatial distribution in the city of Turin; the population distribution; the current energy demand and supply for space heating consumptions;

the buildings connected to the DH network and the ones that could be connected in the future; the distribution of buildings that need renovation; the critical areas of the DH network; the energy performance of buildings after suitable energy retrofit interventions.

2 2. Methods

Starting from a previous study that proposed a methodology to evaluate the potential user connections in the existing DH network for the city of Turin (the goal of the model was to guarantee the maximum amount of new connections), the results shown that the average maximum quota of additional users was +25%, taking into account the territorial, economic and technical constraints of the DH network. The existing DH network was considered and the maximum thermal flux was evaluated in each pipeline fixing the temperatures gradient and the velocity limits. In this paper, to optimize the district heating network operation, a massive retrofit of connected and connectable buildings was hypnotized in order to reduce the energy required in critical pipelines increasing the DH network flexibility. This allowed a significant reduction of the additional mass flow rate of connectable buildings and thus increasing the percentage of connectable buildings. Then, the purpose of this work was to connect all the residential buildings to the existing DH network also through energy efficiency measures. In Fig. 2 the main four phases of the procedure are represented (phase 1 considers the results of previous researches [14–16, 20]; phases 2, 3, and 4 describe the new methodology):

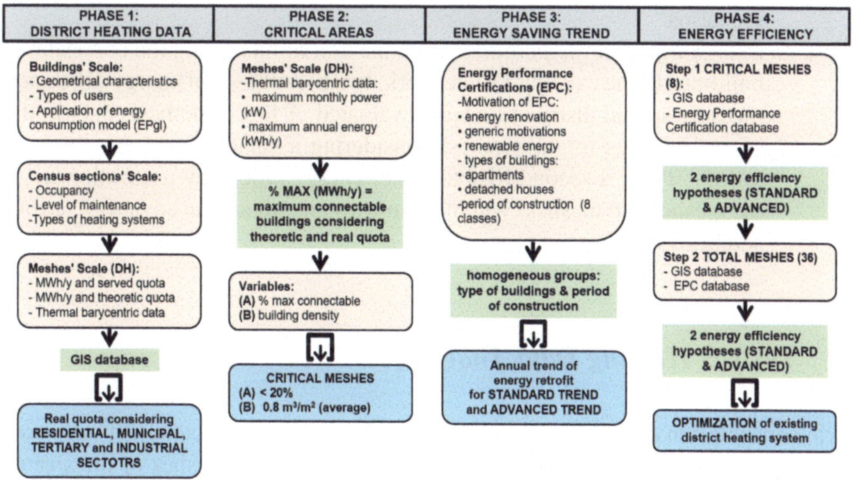

Fig. 2 Flow chart of the procedure

Phase 1. *District Heating data analysis*. With ESRI ArcGIS 10.5.1 software, for 36 meshes with a size of 1 km^2, the DH network barycenters have been localized with their respective heat flows. For each barycenter, a maximum quota of thermal flux was defined [kW; MWh/year] as the maximum heat power that can be supplied by each pipeline of the network considering all territorial, economic and technical constrains. For the 36 meshes, the real connectable quota of buildings [MWh/year] was calculated as the maximum quota minus the already connected thermal load. The connectable quota was evaluated as theoretic and real one: theoretic (corresponds to all connectable buildings and it is higher than the real quota,) considers only territorial and economic constrains, while the real quota considers also technical constrains of pipelines.

Phase 2. *Identification of critical areas*. The most critical meshes have been identified by evaluating the ratio between the real connectable quota and the theoretic connectable quota of buildings; the optimum result should be the connection to the DH network of high percentage of connectable buildings, then critical meshes have a real connectable percentage less than 20% and require urgent retrofit.

Phase 3. *Energy saving trend*. For this analysis, the data about residential buildings' energy performance certificates (EPC) have been used [18]. The available data have been classified considering: two types of buildings (apartments or detached houses); eight classes for the period of constructions (<1918, 1919–1945, 1946–1960, 1961–1970, 1971–1980, 1981–1990, 1991–2005, 2006>); the reasons for the release of EPCs, distinguishing generic and the retrofit interventions. Analyzing the data of the EPCs, the annual standard and advanced energy savings trend were assessed with two future scenarios: standard and advanced energy savings scenarios.

Phase 4. *Energy Efficiency hypotheses*. With the aim of increasing the connectable buildings to the existing DH network, two hypotheses of energy efficiency measure spatial distribution were evaluated. A first hypothesis only in the critical meshes (eight meshes) considering a standard and an advanced retrofit. Then, a second hypothesis of energy efficiency renovation in all the meshes (36 meshes) with the same a standard and an advanced retrofit solutions.

2.1 District Heating Data Analysis

Turin consumption data [MWh/year] and heated volumes connected [m^3] were provided by the DH company. Data were provided through 36 meshes of the size of 1 km × 1 km for in the South-West part of the city of Turin. With a GIS tool, 177 thermal barycenters of the DH network were georeferenced, and for 36 meshes of 1 km^2 the thermal barycenters data have been associated. The connectable buildings to the DH network considering four types of user (residential, municipal, tertiary, industrial) [14] have been identified. In particular, in the GIS database there were information concerning: the quota of energy consumption of connected buildings [MWh/year], and the theoretic connectable quota of buildings and the relative thermal energy consumptions [MWh/year].

Using the results obtained from the fluid dynamic analysis of the DH network, the real amount of connectable buildings has been evaluated [15] and the GIS database has been implemented. Then, for each mesh the district heating data was evaluated, considering the maximum, the connected and the connectable potential quota, the type of users, the location (microclimatic conditions), and the buildings characteristics.

2.2 Identification of Critical Areas

An evaluation of the District Heating characteristics was made to identify the critical areas with a priority of retrofit interventions. With the use of the fluid-dynamic model, values of real thermal energy flows [kW] were provided. Through the following procedure, the maximum percentage of connectable buildings to the existing DH network was calculated:

$$\text{max thermal flow} = \text{connected thermal flow} + \text{real connectable thermal flow}_{st} \quad (1)$$

$$\text{real connectable thermal flow}_{st} < \text{theoretic connectable thermal flow}_{st} \quad (2)$$

where:

- *max thermal flow [MWh/year]*: maximum thermal power that can be supplied by the existing network;
- *connected thermal flow [MWh/year]*: power supplied by the network for the already connected buildings;
- *real connectable thermal flow [MWh/year]*: additional power that can be supplied by the existing network (considering also the limits due to the fluid-dynamic model of DH network);
- *theoretic connectable thermal flow [MWh/year]*: additional power that can be supplied by the existing network (considering only territorial and economic limits);
- *st*: "standard" without retrofit interventions.

Then, the critical meshes have been identified where it is possible to connect less than 20% of the connectable buildings.

2.3 Energy Saving Trend

The energy saving trend of the residential sector has been evaluated, considering the available data on building renovations derived from the energy performance certificates (EPC) from September 2009 to 2015. For this analysis, disregarding the first partial and anomalous years, only EPCs of 2011–2015 were considered. Homogeneous groups of EPCs have been identified, considering: the type of buildings, the period of construction and the motivation for the release of the EPCs. For each group, the annual energy performance, the energy savings and the reachable energy performance [kWh/m^2/year] were calculated. The energy performance was evaluated from the EP_{gl} index considering the energy consumptions for space heating and hot water production.

Subsequently, two future scenarios for standard and advanced retrofits were hypothesized applying the energy saving models. The current annual energy saving trends was also evaluate. This procedure was based on median values of energy intensities for each type of buildings and homogeneous group of EPCs.

2.4 Energy Efficiency Hypotheses

A priority of retrofit interventions has been evaluated, initially with the hypothesis to intervene only in the most critical areas of Turin (identified in phase 2) and then on the whole urban area.

1. Energy efficiency interventions in the most critical areas, that are the meshes in which it is not possible to connect to the existing network more than 20% of the theoretic quota; evaluation of the energy savings achievable with the retrofit of the:
    (a) *the connectable buildings*
    (b) *the connectable buildings and the connected portion.*
2. Energy efficiency measures in the whole city of Turin; evaluation of the energy savings achievable with the retrofit of:
    (a) *the connectable buildings*
    (b) *the connectable buildings* and *the connected portion.*

Through the following procedure, the connectable buildings quota was calculated using the energy savings models with the existing energy policies (standard) and with more deep retrofit interventions (advanced). The current annual energy savings was calculated deducting the energy intensities of EPCs released for generic interventions and the ones of EPCs for retrofit interventions. The advanced annual energy savings instead was calculated deducting the energy intensities of EPCs released for retrofit interventions and the reachable energy intensities.

In particular, comparative analysis of the energy assessment results of the residential area before and after the application of sustainable retrofitting strategies have been made, calculating the improvement potential with the real building stock. The results of the energy assessment of residential buildings were uploaded with GIS, with the spatial distribution of energy consumptions.

3 Case Study

The energy efficiency model was applied to the city of Turin, then the different retrofit measures and the reduction of CO_2 emissions were analysed. To promote the potential expansion of DH network, a study on the optimization of the heat flow distribution has been provided. This analysis could be useful to support a sustainable urban energy plan.

3.1 Characteristics of Turin's Buildings

Turin has about 60,000 of buildings equal to 232 Mm^3, considering the four different types of users:

- residential sector has more than 45,000 of buildings, equal to 164 Mm^3 (70%),
- municipal sector consists in 6000 of buildings, equal to 23 Mm^3 (10%),
- tertiary sector is constituted by 4000 of buildings, equal to 20 Mm^3 (9%) and
- industrial sector with more than 4000 of buildings, equal to 25 Mm^3 (11%).

The residential buildings of Turin have an average surface to volume ratio S/V equal to 0.63 m^{-1}, with mainly big and compact condominiums ($S/V \leq 0.45$ m^{-1}). Although, most of the buildings were built before 1960 (57%) and the 62% have a good level of maintenance (26% excellent and only 11% are in mediocre condition). Considering the existing district heating network, the theoretic connectable buildings consist of 43,000 heated buildings equal to 196 Mm^3. To identify the environmental limits, the localization (hills, Po river, historical center and LTZ) and the type of heating systems (centralized and autonomous systems) have been considered [14]. The residential sector is the most critical for energy consumption analyses; it has almost 32,000 of heated buildings equal to 136 Mm^3 (70%). The municipal, the tertiary, and the industrial sectors have 11,000 heated buildings, specifically 4000 (17 Mm^3), 3000 (18 Mm^3), and 4000 (25 Mm^3).

In this study, energy consumption data were provided on 36 meshes of 1 km × 1 km on the urban territory (with monthly period). The 36 meshes, used in this study, are represented in Fig. 3 in the South-West part of the city with 20,000 of heated buildings equal to 105 Mm^3 and the residential sector represents a significant share, with 16,000 of heated buildings (almost 80 Mm^3). The residential sector accounts for 76%, the municipal sector is 10%, the tertiary sector is 9%, and the industrial sector is 5%.

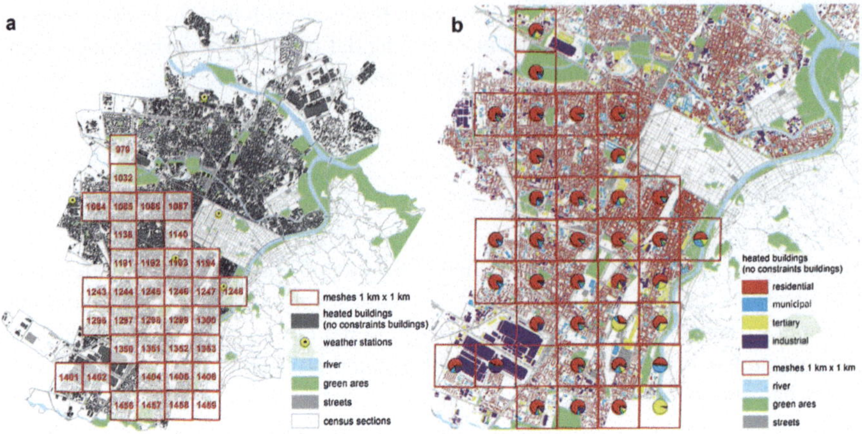

Fig. 3 (**a**) The 36 square meshes of 1 km^2 with ID code (red), weather stations (yellow) and connectable buildings to the DH network; (**b**) Urban distribution of types of users for the 36 meshes: residential, municipal, tertiary, and industrial

In this area, residential buildings are mainly large condominiums, with an average surface to volume ratio equal to 0.52 m^{-1}. The prevailing level of maintenance is good, specifically 32% of buildings have an excellent level, 43% good, 19% mediocre, and 6% bad. Most of the buildings were built before 1960 (53%) and only 21% of the residential buildings were built after 1970, so this buildings stock can represent the general characteristics of the building heritage in the whole city.

The characteristics of buildings in the 36 meshes of Turin (number buildings, heated volume, surface to volume factor, occupancy, types of users, period of construction, main level of conservation, weather station, space heating energy consumptions) have been implemented with a GIS tool to all the analysed buildings with the Municipal Technical Map of Turin (2015) and the Census Data (2011). In particular, the following average characteristics have been calculated for each mesh: the number of buildings, the main period of construction, the surface to volume ratio $S/V_{(avg)}$, the heated volumes, the types of buildings (residential or nonresidential), the main level of maintenance and the level of the apartments used. Data on energy space heating consumption (EP_{gl}) were evaluated by the use of simplified energy models at urban scale for the city of Turin [16] considering also the urban heat island (UHI) phenomenon that influences the microclimate with significantly higher temperatures in the city center and the lowest temperatures in the suburbs and in the surrounding rural and hilly areas [21]. To take into account the UHI effect and its influence on space heating consumptions, different microclimate characteristics were analysed considering the nearest weather station (Fig. 3a) on the average heating season of 2011/2012 [16]. In Fig. 3a also the territorial constrains are shown with the hills on the South-East part of the city across the Po river and the historical center. Figure 3b represents the types of user considered in the simplified energy models at urban scale for the evaluation of the energy consumptions.

3.2 *Characteristics of Turin's District Heating Network*

The Turin district heating network serves about half of the users located in the urban area, for a total value of about 60 Mm^3 of buildings. In particular, for the analyzed 36 meshes, 51 Mm^3 of buildings were connected to the DH network corresponding to more than 48% of users. Furthermore the potentially theoretic connectable buildings are 42 Mm^3 (40% of users), but the real potential expansion allows the connection of only 25% of users taking into account also the technical limits of the pipelines (i.e., maximum mass flow or velocity).

In Table 1 the data of the potential expansion of the DH network are shown for all the 36 analyzed meshes; in particular there were reported: the data of the buildings

Table 1 Connected portion, connectable buildings of DH network and additional deliverable power/consumption: heated volume (% and m^3) and thermal energy consumptions (MWh/year)

ID mesh	Connected quota		Connected thermal consumption	Connectable buildings theoretic quota		Connectable buildings theoretic consumption	Additional deliverable power (real quota)	Additional deliverable consumption (real quota)
	[%]	[m³]	[MWh/y]	[%]	[m³]	[MWh/y]	[MW]	[MWh/y]
979	49.09	586,681	51,315	33.28	397,748	17,080	52	30,783
1032	22.77	189,027	38,581	57.53	477,595	22,197	6	3,327
1084	21.75	593,715	127,303	53.73	1,467,010	68,400	7	4,007
1085	29.22	1,224,495	183,795	56.42	2,363,999	103,690	13	7,505
1086	26.83	1,160,967	186,026	52.69	2,279,739	98,017	20	11,680
1087	32.06	1,886,912	246,982	50.27	2,958,534	124,160	20	11,899
1138	64.88	3,026,070	212,264	28.76	1,341,488	61,051	58	34,291
1140	54.00	2,008,939	140,404	35.45	1,318,749	49,773	50	29,706
1191	37.82	876,846	108,339	42.74	990,957	46,305	15	9,065
1192	34.99	1,346,974	169,325	42.10	1,620,288	71,278	22	12,774
1193	56.36	2,874,301	198,200	36.85	1,879,377	65,969	69	40,376
1194	31.63	1,393,934	184,482	45.84	2,020,053	84,567	12	6,865
1243	74.42	790,217	50,322	17.19	182,493	8,649	7	4,120
1244	62.78	2,057,371	151,878	28.01	917,834	42,540	37	21,655
1245	47.88	1,575,241	139,346	41.87	1,377,608	58,346	30	17,729
1246	53.29	2,095,972	164,742	40.43	1,589,902	66,598	36	21,334
1247	18.18	969,026	228,418	67.17	3,579,794	153,436	10	5,875
1248	2.26	49,000	75,811	72.98	1,581,516	55,327	0	159
1296	100.0	2,676,882	114,403	0	0	0	32	18,623
1297	61.27	2,354,104	173,446	35.35	1,358,196	61,312	42	24,996
1298	84.29	2,915,578	145,110	13.91	481,125	20,184	49	29,003
1299	43.12	1,116,560	123,083	48.99	1,268,435	60,294	18	10,566
1300	63.98	2,967,762	208,930	28.19	1,307,837	58,907	52	30,822
1350	49.46	994,865	104,846	39.57	795,867	41,485	10	6,182
1351	59.84	2,112,957	153,030	32.93	1,162,573	50,386	56	33,109
1352	81.12	2,350,986	88,239	17.14	496,799	15,127	12	7,354
1353	75.18	2,026,351	119,412	18.60	501,258	22,209	41	24,366
1401	36.08	372,317	53,020	40.78	420,808	21,620	11	6,520
1402	19.70	227,236	78,423	63.17	728,553	49,542	6	3,605
1404	53.55	1,572,648	131,574	37.73	1,108,070	49,643	41	24,102
1405	46.41	1,182,307	105,699	44.48	1,133,150	47,018	29	16,867
1406	100.0	839,804	30,481	0	0	0	47	27,594
1456	46.79	993,775	102,225	41.62%	883,987	42,550	24	13,919
1457	46.49	801,192	85,847	40.13%	691,634	34,450	15	8,549
1458	29.55	500,411	75,537	53.59%	907,489	40,479	20	12,018
1459	3.26	19,778	16,475	77.19%	468,591	12,716	9	5,115

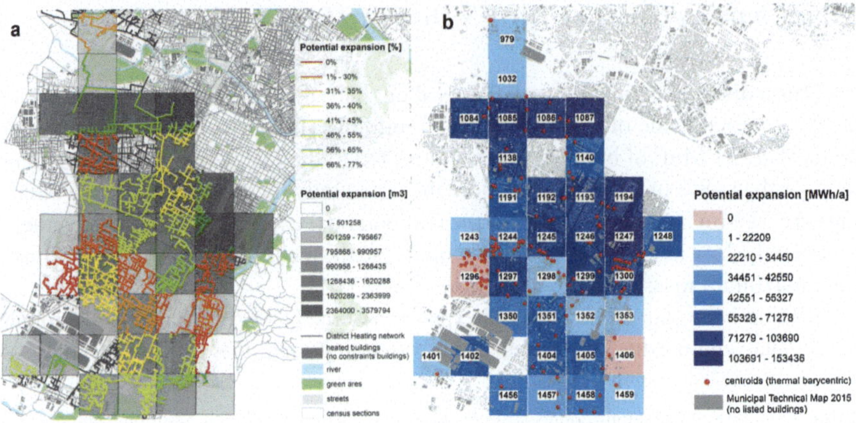

Fig. 4 (**a**) Total potential expansion of the existing DH network: theoretic connectable quota [% and m^3]; (**b**) Total potential expansion of the existing DH network: thermal energy consumptions [MWh/year]

connected to the DH network and the data of theoretic and real connectable quota (heated volumes and their relative thermal energy consumptions). In Table 1, it is possible to observe that two meshes (i.e., 1296 and 1406 in grey) do not have a potential connectable quota because in those areas all connectable buildings are already connected to the existing DH network. In the last columns, the additional deliverable power/consumption (i.e., real connectable quota) was obtained through the knowledge of the design mass flow rates and temperature gap at the primary side of the heat exchangers installed in the substations knowing the already connected quota. Because the delivery temperature at the thermal plant is fixed at 120 °C and assuming a temperature gradient due to losses of 3 °C, a temperature gap of 50 °C was assumed (taking into account that these temperatures can vary a lot during the year).

In Fig. 4 the 36 meshes were represented with the information of the total potential expansion considering the theoretic connectable quota [% and m^3] and the relative thermal energy consumptions [MWh/year]. Figure 4a shows that areas with low buildings densities are already saturated and then with low potential expansion of the existing DH network. On the other hand, in Fig. 4b it can be observed that in the northern peripheral areas of the network, the higher potential expansion corresponds to the higher buildings densities zones of the city.

3.3 Characteristics Energy Retrofit of Turin's Buildings

In Italy, national incentives with tax deduction have been established from Law 296/2006 (in Italian: "Legge Finanziaria 2007") to improve energy efficiency measures and the use of renewable energy technologies on the existing buildings heritage. These incentives mainly relate to residential buildings that replace the windows (43.8%), remake and isolate the roof (19.7%), and replace the old boiler with a condensing one (16.4%) [22].

The data of Turin's buildings renovations refers to the database of the Energy Performance Certificates (EPC) registered for the city of Turin from 2009. This database of EPCs was divided into two groups according to the types of residential buildings: apartments (type A) and detached houses (type B). For this two types of residential buildings, three more classes have been identified referring to the motivation of the EPC release: generic motivations (purchase, sale and location of building), energy renovation (building retrofit) and new installation of renewable energy technologies (photovoltaic, solar thermal collectors, biomass boiler, etc.). Figure 5a shows the EPC data classified according to the motivation of the EPC; mainly EPCs were released for sale or rent of buildings (generic motivation) and only 4% for building retrofit and renewable energy sources. Figure 5b shows the EPC database with the classification according to the types of residential buildings (Type A apartments, Type B detached houses) and the period of construction. As already mentioned, in a big city like Turin, typical buildings are condominiums built in 1946–1970; only the 2% are detached houses.

For each period of construction (distinguishing apartments and detached houses) the specific average and median annual energy consumptions (kWh/m^2/year) were analyzed for every types of motivation: the specific consumption and the achievable consumption after the energy retrofit.

In Tables 2 and 3, the median values of specific energy consumptions are reported for every period of construction; the median value was chosen to neglect anomalous data too low or too high. In the fourth and fifth columns, the energy performance EP_{gl} of EPCs is reported respectively for generic and for retrofit motivation. In the sixth column, the reachable EP_{gl} of EPCs released for retrofit motivation is reported. These three values consent the evaluation of energy savings after standard and advances retrofit interventions (in the last two columns). In Table 3, it is possible to observe that detached houses have an EP_{gl} +29% higher than condominiums with a double energy savings; however, more data are need for detached houses as the number of EPC for retrofit interventions is too low.

From the database of EPCs it is possible to extract also the number of apartments renovated each year and their area in m^2 (from 2009 to 2015). In Fig. 6a, the surface of the renovated apartments was represented and, with the exception of the first year, this surface is quite constant. From the renovated surfaces and the energy savings reported in Table 2, the energy savings trend could be hypnotized (in Fig. 6b). In particular, Fig. 6b shows the cumulative sum of energy savings after buildings retrofit.

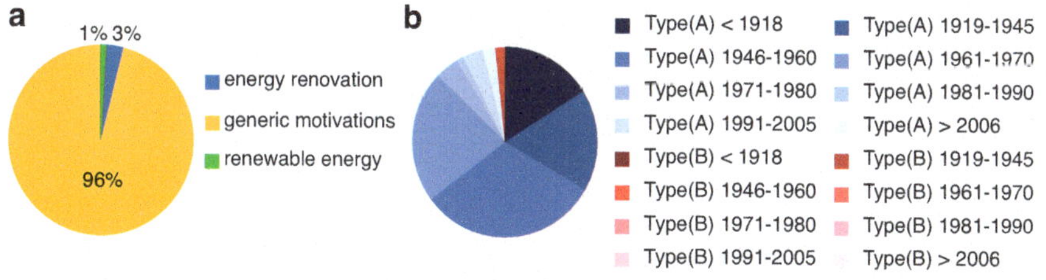

Fig. 5 (**a**) Energy certification motivation: energy retrofit 3%, generic motivations 96% and renewable sources 1%; (**b**) Distribution of types of buildings by period of construction: Type A, apartments (98%) and Type B, detached houses (2%)

Table 2 Characteristics of residential apartments (type A) annual energy savings analysis (EP_{gl} = global Energy Performance, for space heating and domestic hot water)

Period of construction	Number of EPC for generic motivation	Number of EPC for retrofit	EP_{gl} (generic motivation) [kWh/m²/year]	EP_{gl} (standard energy retrofit) [kWh/m²/year]	EP_{gl} (advanced energy retrofit) [kWh/m²/year]	Standard energy savings [kWh/m²/year]	Advanced energy savings [kWh/m²/year]
<1918	16,431	657	213	165	127	48	86
1919–1945	17,495	565	224	165	126	59	98
1946–1960	31,961	1004	213	163	125	49	87
1961–1970	23,109	658	209	175	107	33	102
1971–1980	4371	79	200	172	120	29	80
1981–1990	1339	28	179	136	114	44	65
1991–2005	3954	66	132	121	110	11	22
2006>	2000	25	96	87	80	0	15

Table 3 Characteristics of residential detached houses (type B) annual energy savings analysis

Period of construction	Number of EPC for generic motivation	Number of EPC for retrofit	EP_{gl} (generic motivation) [kWh/m^2/year]	EP_{gl} (standard energy retrofit) [kWh/m^2/year]	EP_{gl} (advanced energy retrofit) [kWh/m^2/year]	Standard energy savings [kWh/m^2/year]	Advanced energy savings [kWh/m^2/year]
<1918	303	39	275	183	132	92	143
1919–1945	374	29	335	188	120	147	215
1946–1960	478	68	291	170	125	121	166
1961–1970	239	33	290	205	157	85	133
1971–1980	82	9	243	206	132	37	111
1981–1990	67	17	200	166	102	34	98
1991–2005	62	5	198	167	104	31	94
2006>	40	6	89	94	70	0	19

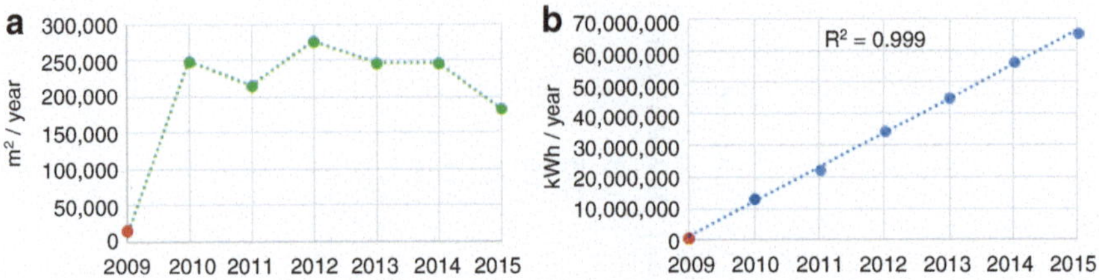

Fig. 6 (**a**) Surface of retrofit interventions: retrofit quota of residential apartments (type A) from September 2009 to December 2015; (**b**) Energy savings after retrofit interventions: energy savings trend of residential apartments (type A) from September 2009 to December 2015

4 Results and Discussion

For each mesh, the characteristics of the residential buildings were analyzed by collecting data and classifying the buildings according to the period of construction and building typology with a GIS tool (Fig. 7a). In the historical center of the city, buildings were built mainly before 1945, while in the other parts, buildings were mainly built in 1961–1980. Only in three meshes, there is a high percentage of buildings built in 1981–1990.

For each homogeneous group of buildings the following parameters have been calculated: the number of buildings, heated area [m^2], heated volume [m^3], specific consumption [kWh/m^3/year], annual consumption [MWh/year], connected portion, theoretic connectable portion, and real connectable portion [MWh/year].

Considering the load losses in the DH network and the energy consumption of the connected buildings, the real connectable quota of residential buildings for each thermal barycenter and mesh was evaluated. Figure 7b shows the real quota of residential buildings that can be connected to the existing DH network; an expansion of 0% indicates that the meshes are saturated with all potential residential buildings already connected. The last mesh in the lower right corner is saturated as no residential building is still connectable.

As we have seen, the expansion of the DH network also depends on the number of buildings present in the meshes and therefore on their relative buildings density. Therefore, in the evaluation of the critical meshes (with real connectable quota <20% than the theoretic quota and excluding the saturated meshes), the average building density of these eight meshes (in Fig. 7b) was calculated: 0.7 m^3/m^2.

Figure 8b shows that the critical meshes are located in peripheral areas where the network is less dense. The building density is greater than 0.85 m^3/m^2 for six critical meshes, on the other hand two meshes (1032 and 1248) have a very low density (0.2–0.25 m^3/m^2) but are nevertheless considered critical because in those areas the potential expansion is less than 20%.

Figure 8b also shows that there are two saturated meshes, 1296 and 1406. For these two meshes the connectable quotas, equal to 18,622 MWh/year and 27,593 MWh/year, are considered as surplus and the supplied energy could be used to serve other buildings. In addition, the mesh no. 1459 has a potential quota

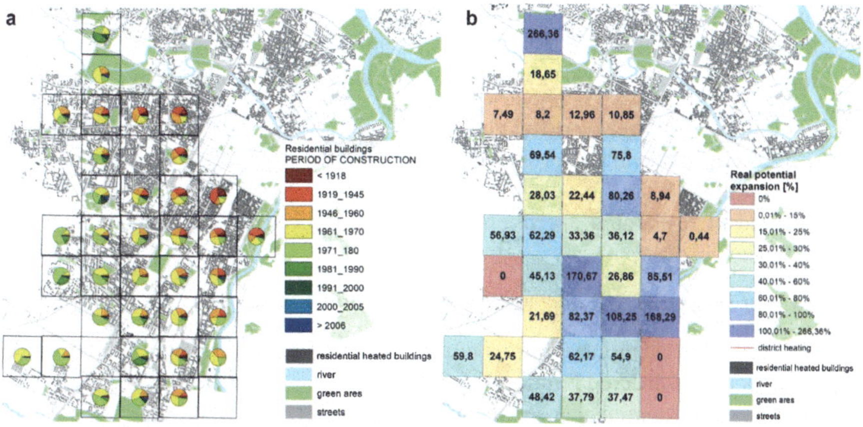

Fig. 7 (**a**) Distribution of residential buildings considering the period of construction; (**b**) Real connectable buildings with a potential expansion of the existing DH network (considering the residential sector)

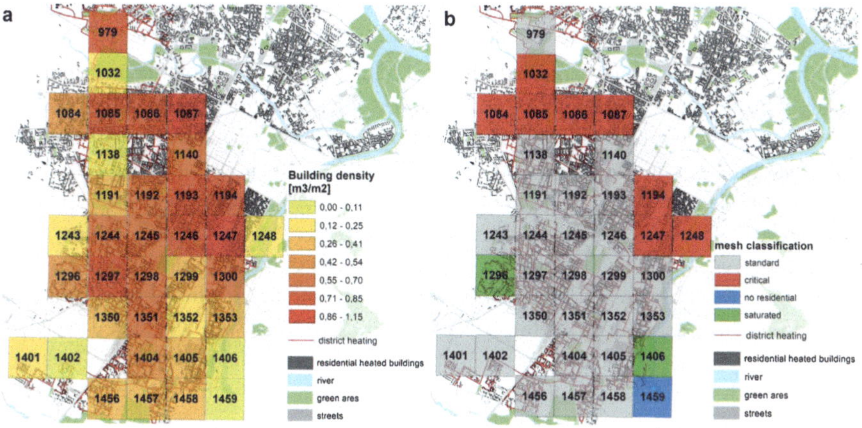

Fig. 8 (**a**) Distribution of residential building density; (**b**) Classification of 36 meshes: standard, critical, nonresidential buildings, saturated

(5115 MWh/year) that could be used to serve nonresidential buildings because in this area there are tertiary and industrial buildings (with an energy required of 12,716 MWh/year).

With the application of the energy savings models, two future group of scenarios were assumed considering a standard and advanced building retrofit. For the first group of scenarios, only a retrofit of connectable residential buildings was proposed; while the second group of scenarios presented a retrofit for all buildings (connectable plus connected residential buildings). Moreover, these scenarios were also subdivided in more scenarios with these interventions only on the critical meshes (*phase one*) and on the whole area (*phase two*).

The scenarios analyzed, that evaluate the energy savings (ES) through the hypotheses of energy efficiency, are the following:

Group of scenarios 1. Energy savings (ES) assessment considering:

1a. ES trend of residential apartments without other energy retrofit interventions;
1b. ES trend of residential apartments plus a standard retrofit for the real connectable buildings.
1c. ES trend of residential apartments plus advanced retrofit for real connectable buildings.

Group of scenarios 2. Energy saving assessment considering:

2a. ES trend of residential apartments plus standard retrofit for real connectable buildings and connected buildings.
2b. ES trend of residential apartments plus advanced retrofit for the real connectable buildings and connected buildings.

In the first phase, the energy savings hypotheses were evaluated only for the most critical meshes and the buildings that need an urgent retrofit have been identified in order to optimize the heat distribution in the DH network. Then, energy savings of residential buildings were evaluated considering the eight critical meshes represented in Fig. 8b.

Table 4 shows for each scenario the average real connectable quotas of buildings in the eight critical meshes [%, MWh/year] and the energy savings [MWh/year] taking into account standard and advanced retrofit applied to connectable and connected buildings. Intervening only on the critical meshes there is a moderate energy savings; in particular, the real connectable quota in the first scenarios varies from 8.5% in scenario 1a to 16.3% with advanced retrofit in scenario 1c. In scenarios 2, intervening with retrofit measures on all buildings (connected and connectable) the percentage of real connectable quota reaches 57.9% (scenario 2b), with the advanced retrofit.

In *phase two*, the energy savings of residential buildings were evaluated considering the whole district heating area (36 meshes). In this case, the surplus of energy obtained with retrofit interventions is significant, especially with the advanced retrofit measures applied to all buildings (scenario 2b).

In Table 5 the results of phase 2 are shown with the surplus quota in brackets. The results shown a considerable reduction of thermal energy consumptions, with the

Table 4 Energy savings assessment of Turin critical areas (eight meshes, phase 1)

Number meshes	Scenario	Average real connectable quotas in the eight meshes [%]	Theoretic connectable quota [MWh/year]	Real connectable quota [MWh/year]	Energy savings [MWh/year]
8	1a	8.5	601,012	51,318	0
	1b	11.1		66,925	139,482
	1c	16.3		98,153	286,886
	2a	25.1		151,010	203,457
	2b	57.9		348,240	416,897

Table 5 Energy savings (ES) assessment of whole Turin areas (36 meshes, phase 2)

Number meshes	Scenario	Average of real connectable quotas (global with surplus) [%]	Theoretic connectable quota [MWh/year]	Real connectable quota [MWh/year]	Energy savings (ES) [MWh/year]	Surplus (energy savings, ES) [MWh/year]	Maximum energy required [MWh/year]
36	1a	34.3 (40.5)	1,410,985	483,446	0	87,899	2,169,097
	1b	41.5 (49.7)		585,557	294,786	116,126	2,140,870
	1c	52.5 (71.3)		741,369	650,375	264,897	1,992,099
	2a	57.8 (79.0)		815,559	626,842	298,579	1,958,317
	2b	81.7 (167.6)		1,152,841	1,407,373	1,211,706	1,045,290

possibility to connect almost all the potential buildings. In particular, in scenario 1 it is possible to connect 34.3% (40.5%) of potential residential buildings without energy retrofit (1a) and 53% (71%) with an advanced retrofit of the residential building stock (1c). In scenario 2, there is a connectable quota equal to 58% (79%) of the total through a standard retrofit, but with scenario 2b it is possible to reach 82% (168%).

The maximum energy supplied for the whole DH area is equal to 2,256,996 MWh/year, of which 1,680,536 MWh/year are connected buildings and 87,899 is a surplus. Taking into account the energy savings for the different scenarios, the energy supplied can be considerably reduced from 2,169,097 MWh/year (scenario 1a) to 1,045,290 MWh/year (scenario 2b). The energy required through an advanced retrofit of all residential buildings is reduced by 52%.

Figures 9 shows the maximum energy required for the different scenarios. For each mesh the energy quota required to cover the energy consumption was indicated

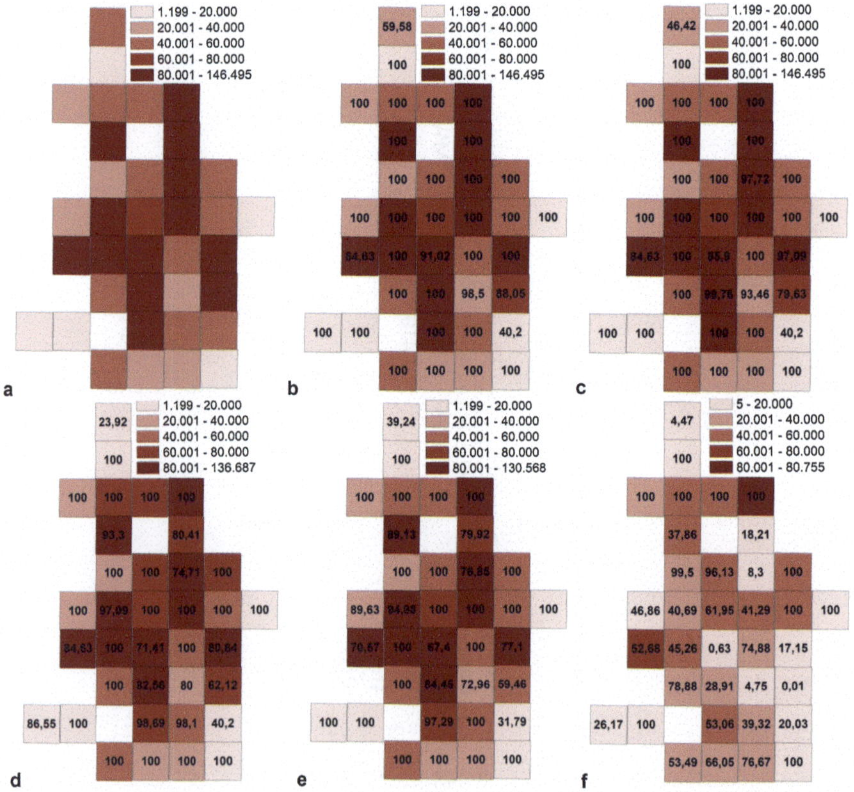

Fig. 9 Maximum energy supplied [MWh/year] for: (**a**) case study; (**b**) 1a, energy savings (ES) without energy retrofit; (**c**) 1b, ES with standard retrofit of residential connectable buildings; (**d**) 1c, ES with advanced retrofit of residential connectable buildings; (**e**) 2a, ES with standard retrofit of residential connectable and connected buildings; (**f**) 2b, ES with an advanced retrofit of residential connectable and connected buildings

(% and MWh/year) considering the current situation (scenario 1a), the standard retrofit (scenarios 1b and 2a) and the advanced retrofit (scenarios 1c and 2b).

The retrofitting measures lead to a change in the water dynamic within the network. The evaluation of the effects on the water velocities within the pipelines is here performed through a fluid-dynamic model of the district heating network. The model is based on the application of mass and momentum to respectively the nodes and the branches of the network; the network topology is expressed through a graph approach. The model allows evaluating the pressure and the mass flow rates in all the nodes and branches of the network. It has been validated [23, 24] and it has already been used with the aim of pumping cost reduction [25] and thermal peak shaving [26–28]. All the scenarios analyzed in this study (1a, 1b, 1c, 2a, and 2b) are simulated through the fluid-dynamic model and the results obtained are shown in Fig. 10. In the figure, the velocity in all the branches is plotted for all the scenarios considered. The velocity of the water is always lower than 4.5 m/s, thus acceptable. The velocity values tent to decrease while the retrofitting measure become more significant, that is, the energy required by the user decreases. In case 2b, which is the scenario presenting the most effective retrofitting measures, the higher velocity value is about 3 m/s. The values progressively increases until 4.5 m/s in case 1a. The mean velocity value in the network also decreases, with increased weight of the retrofitting measures; in particular, the scenario 1a presents a mean velocity of 1.4 m/s, while the value is 0.9 m/s in the scenario 2b.

Then, the results show that in future scenarios with retrofit interventions there will be a decrease of the energy consumption with an improvement in the operation of the DH network and an over energetic potential (surplus) that could be used to serve other nearby areas; the reduction of energy consumption will determine also

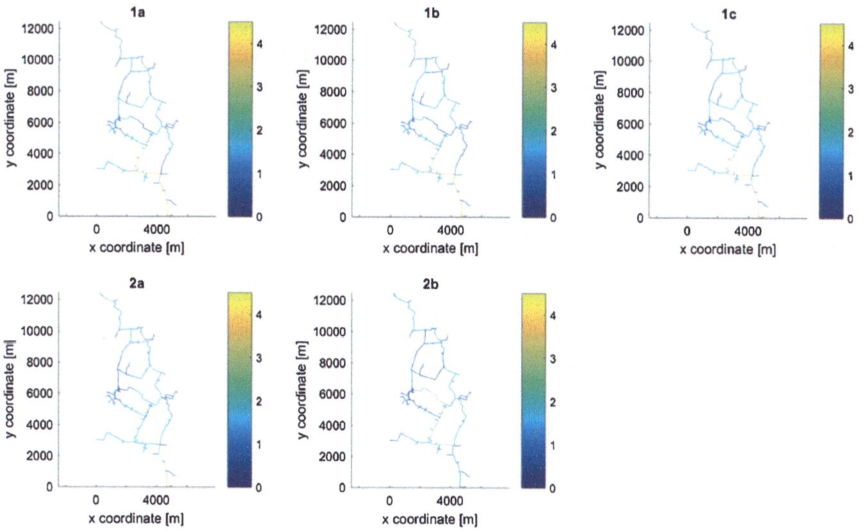

Fig. 10 Comparison of velocity values [m/s] in the branches of the network between standard and advanced retrofit

Table 6 Greenhouse gas emissions for each scenario

Number meshes	Scenario	Energy consumptions (DH) [MWh/year]	Energy consumptions (centralized systems) [MWh/year]	GHG emissions (DH) [tonCO_2]	GHG emissions (centralized systems) [tonCO_2]	GHG emissions (autonomous systems) [tonCO_2]	Total emissions [tonCO_2]
36	0	1,680,536	1,410,985	257,122	329,270	90,155	676,547
	1a	2,169,097	927,471	331,872	216,436		638,463
	1b	2,140,870	825,024	327,553	192,529		610,237
	1c	1,992,099	669,848	304,791	156,317		551,263
	2a	1,958,317	595,204	299,622	138,898		528,675
	2b	1,045,290	257,896	159,929	60,183		310,267

a consequent decrease of greenhouse gas emissions and the possibility to connect more users with therefore an increase of the livability or urban spaces and probably an improvement of social life.

Considering the energy consumptions and the mix of the energy sources, the greenhouse gas emissions were also evaluate (in Table 6) in tonCO_2/MWh [29]: gas oil = 0.280, LPG = 0.240, carbon = 0.370, fuel oil = 0.290, electricity energy = 0.460, natural gas = 0.210, solar thermal = 0.0; for the DH, the results of Data4Action project for Turin with 0.153 tonCO_2/MWh. In the 36 meshes connected to the DH network, the buildings with centralized heating system are 78%, of which most of the buildings use natural gas (79.1%), a small portion uses gas oil (14.6%) and electricity energy (5%). Buildings with autonomous heating systems (22%) also use mainly natural gas (91.1%) and electricity energy (4.6%). Then, most of the residential buildings use natural gas for space heating and the distribution of the use of energy sources is consistent with the national residential energy consumption by source [30]. The existing emissions, considering the current state, amounted to 676.547 tonCO_2, then the emissions decrease by 19% with an advanced retrofit only for the real connectable buildings (scenario 1c) and they decrease up to 54% with an advanced retrofit for the real connectable and connected buildings (scenario 2b).

This study is useful for future hypotheses of energy efficiency policies, to understand how to exploit the energy surplus achieved thanks to the retrofit of the residential sector and to evaluate a possible retrofit of other sectors (municipal, tertiary, and industrial). Furthermore, the socioeconomic variables that influence the energy consumption trend could be taken into consideration, such as the characteristics of the inhabitants and the income [31] and engineering models at district scale will be developed to take into account also the hourly energy profiles [32].

5 Conclusion

To achieve energy sustainability in an urban context such as Turin, with high building density, it is necessary to optimize: the energy consumptions of different users, the distribution of heat through the DH network but also the use the buildings envelope and the urban surfaces to produce energy from renewable sources. The limited availability of renewable energy sources in urban context, lead to combine these strategies with energy savings measures to reduce the use of energy in buildings.

The methodology presented in this work offers a global GIS-based assessment at urban scale to evaluate the energy consumption spatial distribution, the potential expansion of the DH network and the future scenarios of building retrofit hypotheses for a more sustainable city. This analysis will allow to evaluate the results of current and future energy policies and to identify the more effective actions adapted to the real territorial and socioeconomic context. Besides, the local administration could promote the DH network expansion with the renovation of buildings in critical areas, to increase the quota of connected DH users. From the energy retrofit of buildings, there are advantages at an economic and environmental point of view, and with the optimization of the district heating network it is possible to connect more users, in order to reduce polluting emissions and to reduce energy consumption.

The use of GIS tools is essential in order to overlay different data, to perform multilayer analyses and to replicate the proposed assessment in other territories.

References

1. Ang, B. W., Choong, W. L., & Ng, T. S. (2015). Energy security: Definitions, dimensions and indexes. *Renewable and Sustainable Energy Reviews, 42*, 1077–1093. https://doi.org/10.1016/j.rser.2014.10.064.
2. UNEP. (2016). *The emissions gap report*. Retrieved March, 2018, from http://orbit.dtu.dk/files/127120606/EGR_2016_1_.pdf.
3. Delmastro, C., Mutani, G., & Corgnati, S. (2016). A supporting method for selecting cost-optimal energy retrofit policies for residential buildings at the urban scale. *Energy Policy, 99*, 42–56. https://doi.org/10.1016/j.enpol.2016.09.051.
4. Caputo, P., & Pasetti, G. (2015). Overcoming the inertia of building energy retrofit at municipal level: The Italian challenge. *Sustainable Cities and Society, 15*, 120–134. https://doi.org/10.1016/j.scs.2015.01.001.
5. Hervás, M., Sendra, J., & Suárez, R. (2017). Towards an energy assessment on an urban scale for retrofitting the housing stock in mediterranean cities. *Procedia Environmental Sciences, 38*, 688–695. https://doi.org/10.1016/j.proenv.2017.03.150.
6. Olsthoorn, D., Haghighat, F., & Mirzaei, P. (2016). Integration of storage and renewable energy into district heating systems: A review of modelling and optimization. *Solar Energy, 136*, 49–64. https://doi.org/10.1016/j.solener.2016.06.054.
7. Lund, H., Möller, B., Mathiesen, B. V., & Dyrelund, A. (2010). The role of district heating in future renewable energy systems. *Energy, 35*, 1381–1390. https://doi.org/10.1016/j.energy.2009.11.023.
8. Sperling, K., & Möller, B. (2012). End-use energy savings and district heating expansion in a local renewable energy system - A short-term perspective. *Applied Energy, 92*, 831–842. https://doi.org/10.1016/j.apenergy.2011.08.040.
9. Johansson, T., Olofsson, T., & Mangold, M. (2017). Development of an energy atlas for renovation of the multifamily building stock in Sweden. *Applied Energy, 203*, 723–736. https://doi.org/10.1016/j.apenergy.2017.06.027.
10. Delmastro, C., Mutani, G., Schranz, L., & Vicentini, G. (2015). The role of urban form and socio-economic variables for estimating the building energy savings potential at the urban scale. *International Journal of Heat and Technology, 33*, 91–100. https://doi.org/10.18280/ijht.330412.

11. Nielsen, S., & Möller, B. (2013). GIS based analysis of future district heating potential in Denmark. *Energy, 57*, 458–468. https://doi.org/10.1016/j.energy.2013.05.041.
12. Gils, H., Cofala, J., Wagner, F., & Schöpp, W. (2013). GIS-based assessment of the district heating potential in the USA. *Energy, 58*, 318–329. https://doi.org/10.1016/j.energy.2013.06.028.
13. Rezaie, B., & Rosen, M. A. (2012). District heating and cooling: Review of technology and potential enhancements. *Applied Energy, 93*, 2–10. https://doi.org/10.1016/j.apenergy.2011.04.020.
14. Guelpa, E., Mutani, G., Todeschi, V., & Verda, V. (2017). A feasibility study on the potential expansion of the district heating network of Turin. *Energy Procedia, 122*, 847–852. https://doi.org/10.1016/j.egypro.2017.07.446.
15. Mutani, G., Todeschi, V., Kampf, J, Coors, V., & Fitzky, M. (2018). Building energy consumption modeling at urban scale: Three case studies in Europe for residential buildings. *IEEE International Telecommunications Energy Conference, INTELEC 2018*; Torino (IT), 7-11th October 2018. https://doi.org/10.1109/INTLEC.2018.8612382.
16. Mutani, G., & Todeschi, V. (2017). Space heating models at urban scale for buildings in the city of Turin (Italy). *Energy Procedia, 122*, 841–846. https://doi.org/10.1016/j.egypro.2017.07.445.
17. Chalal, M., Benachir, M., White, M., & Shrahily, R. (2016). Energy planning and forecasting approaches for supporting physical improvement strategies in the building sector: A review. *Renewable and Sustainable Energy Reviews, 64*, 761–776. https://doi.org/10.1016/j.rser.2016.06.040.
18. *Energy Performance Certifications (EPC) - Information System for Energy Performance of Buildings (SIPEE) in Regione Piemonte*. Retrieved January, 2018, from www.sistemapiemonte.it.
19. Caputo, P., & Pasetti, G. (2017). GIS tools towards a renovation of the building heritage, Climamed 2017 – Mediterranean Conference of HVAC; Historical buildings retrofit in the Mediterranean area, 12–13 May 2017, Matera, Italy. *Energy Procedia, 133*, 435–443. https://doi.org/10.1016/j.egypro.2017.09.388.
20. Guelpa, E., Mutani, G., Todeschi, V., & Verda, V. (2018). Reduction of CO_2 emissions in urban areas through optimal expansion of existing district heating networks. *Journal of Cleaner Production, in: Journal of Cleaner Production, 204*, 117–129. https://doi.org/10.1016/j.jclepro.2018.08.272.
21. Mutani, G., & Fiermonte, F. (2016). The urban microclimate and the urban heat island. A model for a sustainable urban planning. In R. Ingaramo & A. Voghera (Eds.), *Topics and methods for urban and landscape design. From the river to the project. Urban and landscape perspectives* (Vol. 19). Cham: Springer International Publishing. https://doi.org/10.1007/978-3-319-51535-9.
22. ENEA. (2017). *Annual report 2017 – 65% Fiscal deductions. Executive summary*. Rome: Italian National Agency for New Technologies, Energy and Sustainable Economic Development. Retrieved March, 2018, from http://www.efficienzaenergetica.enea.it/allegati/Detrazioni-65-executive-summary-2017-en.pdf.
23. Guelpa, E., Toro, C., Sciacovelli, A., Melli, R., Sciubba, E., & Verda, V. (2016). Optimal operation of large district heating networks through fast fluid-dynamic simulation. *Energy, 102*, 586–595. https://doi.org/10.1016/j.energy.2016.02.058.
24. Guelpa, E., Sciacovelli, A., & Verda, V. (2019). Thermo-fluid dynamic model of large district heating networks for the analysis of primary energy savings. *Energy, 184*, 34. https://doi.org/10.1016/j.energy.2017.07.177.
25. Sciacovelli, A., Guelpa, E., & Verda, V. (2013). Pumping cost minimization in an existing district heating network. In *Proceedings of IMECE; November15-21; San Diego, California, USA*. New York, NY: ASME. https://doi.org/10.3390/app7050488.
26. Guelpa, E., Barbero, G., Sciacovelli, A., & Verda, V. (2017). Peak-shaving in district heating systems through optimal management of the thermal request of buildings. *Energy, 137*, 706–714. https://doi.org/10.1016/j.energy.2017.06.107.

27. Verda, V., Guelpa, E., Sciacovelli, A., Acquaviva, A., & Patti, E. (2016). Thermal peak load shaving through users request variations. *International Journal of Thermodynamics, 19*(3), 168–176. https://doi.org/10.5541/ijot.5000175955.
28. Brundu, F. G., Patti, E., Osello, A., Del Giudice, M., Rapetti, N., Krylovskiy, A., Jahn, M., Verda, V., Guelpa, E., Rietto, L., & Acquaviva, A. (2017). IoT software infrastructure for energy management and simulation in smart cities. *IEEE Transactions on Industrial Informatics, 13*(2), 832–840. https://doi.org/10.1109/TII.2016.2627479.
29. Italian National Agency for New Technologies, Energy and Sustainable Economic Development ENEA.*Poteri calorifici inferiori dei combustibili e fattori di emissione della* CO_2. Rome: ENEA. (in Italian). Retrieved March, 2018, from http://www.efficienzaenergetica.enea.it/regioni/siape/poteri-calorifici-inferiori-dei-combustibili-e-fattori-di-emissione-della-co2.
30. International Energy Agency. (2017). *Statistics: Energy efficiency indicators*. Paris: Author. Retrieved March, 2018, from https://www.iea.org/publications/freepublications/publication/EnergyEfficiencyHighlights_2017.PDF.
31. Mutani, G., & Todeschi, V. (2018). Energy resilience, vulnerability and risk in urban spaces. *Journal of Sustainable Development of Energy, Water and Environment Systems, 6*(4), 694–709. https://doi.org/10.13044/j.sdewes.d6.0203.
32. Mutani, G., Todeschi, V., Grisolia, G., & Lucia, U. (2019). Introduction to constructal law analysis for a simplified hourly energy balance model of residential buildings at district scale. *Tecnica Italiana - Italian Journal of Engineering Science, 63*(1), 13–20. https://doi.org/10.18280/ti-ijes.630102.

Index

P. Bertoldi (ed.), *Improving Energy Efficiency in Commercial Buildings and Smart Communities*, Springer Proceedings in Energy,
https://doi.org/10.1007/978-3-030-31459-0

F

The manufacturer's authorised representative in the EU is Springer Nature Customer Service Centre GmbH, Europaplatz 3, 69115 Heidelberg, Germany. If you have any concerns regarding our products, please contact ProductSafety@springernature.com

Printed and bound by CPI Group (UK) Ltd, Croydon, CR0 4YY
30/09/2026
02225458-0010